Gold in the Black Pine Mining District of Idaho

by United States Geological Survey

with an introduction by Kerby Jackson

Introduction

It has been years since the United States Geological Survey released his important publication "Gold In The Black Pine Mining District of Idaho". First released in 1984, this important volume has now been out of print for this days and has been unavailable to the mining community since those days, with the exception of expensive original collector's copies and poorly produced digital editions.

It has often been said that *"gold is where you find it"*, but even beginning prospectors understand that their chances for finding something of value in the earth or in the streams of the Golden West are dramatically increased by going back to those places where gold and other minerals were once mined by our forerunners. Despite this, much of the contemporary information on local mining history that is currently available is mostly a result of mere local folklore and persistent rumors of major strikes, the details and facts of which, have long been distorted. Long gone are the old timers and with them, the days of first hand knowledge of the mines of the area and how they operated. Also long gone are most of their notes, their assay reports, their mine maps and personal scrapbooks, along with most of the surveys and reports that were performed for them by private and government geologists. Even published books such as this one are often retired to the local landfill or backyard burn pile by the descendents of those old timers and disappear at an alarming rate. Despite the fact that we live in the so-called "Information Age" where information is supposedly only the push of a button on a keyboard away, true insight into mining properties remains illusive and hard to come by, even to those of us who seek out this sort of information as if our lives depend upon it. Without this type of information readily available to the average independent miner, there is little hope that our metal mining industry will ever recover.

This important volume and others like it, are being presented in their entirety again, in the hope that the average prospector will no longer stumble through the overgrown hills and the tailing strewn creeks without being well informed enough to have a chance to succeed at his ventures.

Kerby Jackson
Josephine County, Oregon
October 2015

CONTRIBUTIONS TO ECONOMIC GEOLOGY

GOLD IN THE BLACK PINE MINING DISTRICT, SOUTHEAST CASSIA COUNTY, IDAHO

By Bruce T. Brady

ABSTRACT

Base and precious metals in the Black Pine mining district, Black Pine Mountains, Cassia County, Idaho, occur in two types of deposits—deposits of disseminated gold, and vein deposits containing base metals and subordinate amounts of silver and gold. Geologic studies were undertaken in the district to determine the mode of occurrence and genesis of the mineral deposits. In addition to geologic mapping, mineralized areas were examined and limited geochemical sampling and analysis done.

Sedimentary rocks of Late Mississippian to Early Permian age in the Black Pine mining district consist of limestone, silty and sandy limestone, calcareous sandstone and siltstone, quartzite, dolomite, and argillite. Low-grade metamorphism has altered dark pelitic rocks to argillite and caused recrystallization of many of the carbonate rocks. Poorly exposed narrow mafic dikes of probable Tertiary age crop out in a few places in the district. Quaternary silt, sand, and gravel cover part of the area.

Low-angle faults along which younger rocks have been moved over older ones or over rocks of partly equivalent ages are the dominant structures in the Black Pine Mountains. High-angle normal faults are much less prominent. Folds trend principally north and northeast.

The quartz veins are steeply inclined tabular bodies of pronounced differences in thickness, and have isolated pods and stringers containing some tetrahedrite (freibergite), sphalerite, jamesonite, less pyrite, even less cinnabar, and some oxidation products. Production of lead, silver, gold, and copper from the vein deposits amounted to less than 1,000 metric tons.

Disseminated gold occurs in altered Pennsylvanian silty carbonate rocks. Siltstone is common in what appears to be the principal area of altered rock. Siltstone and claystone there are tan, brown, and green with prominent iron-oxide coatings along numerous small fractures.

Gold and pyrite are the principal metallic minerals and appear to be disseminated chiefly in siltstone. Associated gangue minerals include calcite, barite, and quartz. Gold grains are submicrometer in size, as shown by the electron microscope, and are commonly less than $0.5\ \mu m$ (micrometer) in diameter. Samples from the open pit and vicinity also have anomalous concentrations of mercury, arsenic, and antimony. These elements and tungsten occur in anomalous amounts with disseminated gold deposits elsewhere in the Basin and Range province.

The fine-grained gold presumably was deposited at some depth in a hot spring environment. The age of the mineralization is unknown. It certainly is younger than the Permian rocks and also is younger than the major deformation in the area, but that deformation is not closely dated.

INTRODUCTION

Base and precious metals occur in the Black Pine mining district on the southeast side of the Black Pine Mountains in southeast Cassia County, Idaho (fig. 1). Two types of mineral deposits are identified, and these are mostly in Pennsylvanian strata. One type consists of replacement deposits containing base metals and subordinate amounts of silver and gold, and the other consists of deposits of disseminated gold and minor amounts of associated base metals.

The geologic setting and mineralogy of the base-metal deposits that were worked prior to 1930 have been discussed by Anderson (1931). An occurrence of cinnabar was described by McKaskey (1917) and Larsen (1919), and summaries of Larsen's discussion appeared later (Ransome, 1921; Shannon, 1926; and Bailey, 1964). Other descriptions of mineralized rock in the Black Pine district may be found in Varley and others (1919), Ross (1941), Hubbard (1955), and Ross and Savage (1967). French (1975) briefly summarized the occurrence of fine-grained disseminated gold in the area.

This investigation is part of a geologic study of the Black Pine Mountains within the Strevell 15-minute quadrangle mapped by Smith (1982, 1983). Stratigraphic and structural information presented here derives from that project. Smith cooperated in the more detailed fieldwork in parts of the mining district.

GENERAL GEOLOGY

SEDIMENTARY ROCKS

The Black Pine Mountains are composed predominantly of sedimentary and some slightly metamorphosed sedimentary rocks, which range in age from Devonian to Permian. Units in the mining district are Mississippian to Permian (fig. 2). The Upper Mississippian and Lower Pennsylvanian Manning Canyon Shale is composed mostly of dark-gray to black argillite with lesser amounts of siltstone, claystone, and shale, and much less quartzite and limestone. Pennsylvanian units of the Oquirrh Formation consist of limestone, silty and sandy limestone, quartzitic sandstone lenses, and dolomite. A Pennsylvanian and Lower Permian member of the Oquirrh Formation is composed of calcareous sandstone and siltstone and less silty and sandy limestone and quartzite. Quaternary silt, sand, and gravel cover part of the area.

METAMORPHIC ROCKS

Metamorphism in the Black Pine Mountains, of predominantly low rank, is most obvious in the dark pelitic rocks of the Manning Canyon

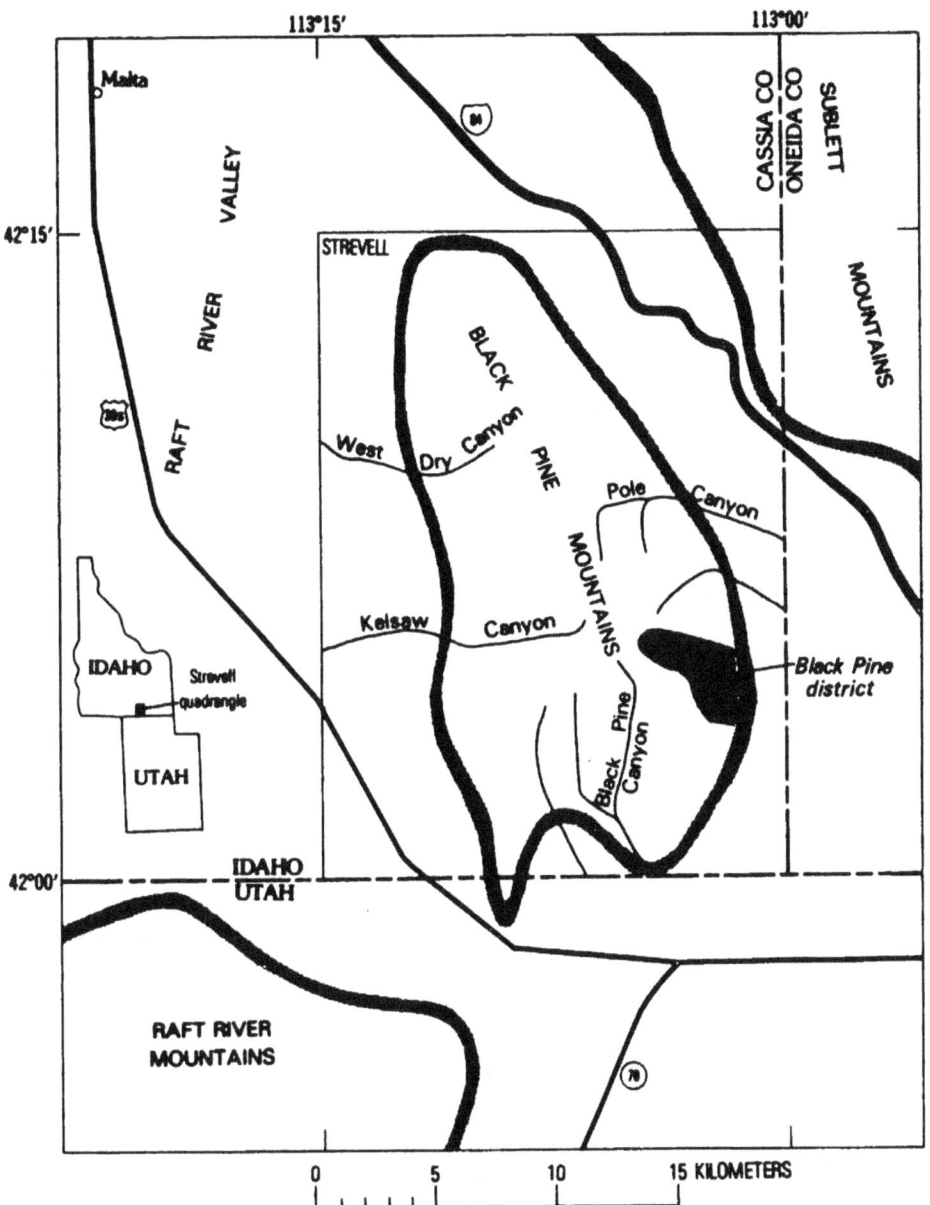

FIGURE 1.—Index map showing location of the Black Pine mining district in the Black Pine Mountains, Cassia County, Idaho.

Shale, which are now chiefly argillites with well-developed fracture cleavage in most places. The mineral assemblages and chemical trends that were produced during metamorphism of the Manning Canyon were discussed in detail by Christensen (1975). These data indicate that temperatures may have reached 350°-400°C during metamorphism. Temperature ranges of 300°-500°C in the area are also indicated by color alteration indices (CAI) of conodonts from Mississippian, Pennsylvanian, and Permian strata (Anita G. Harris, John Repetski,

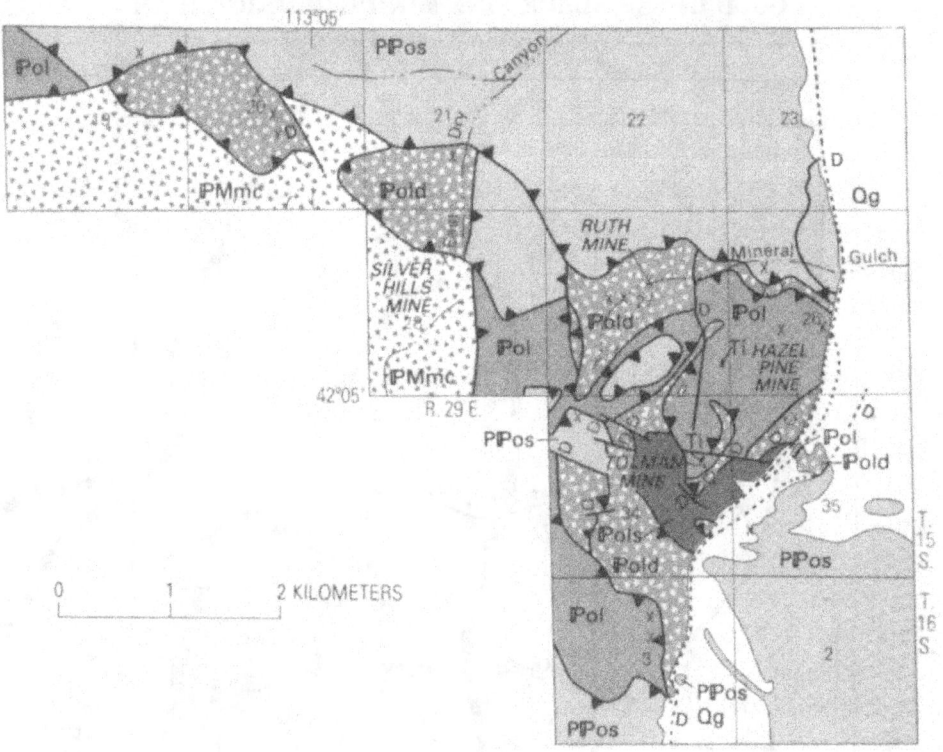

EXPLANATION

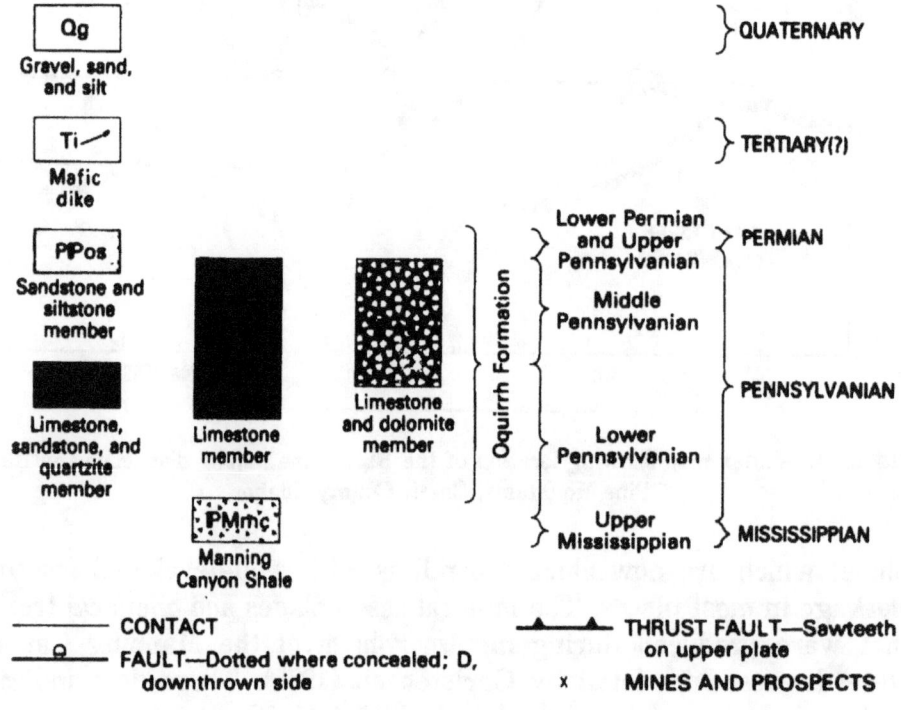

FIGURE 2.—Generalized geologic map of the Black Pine mining district, Black Pine Mountains, Cassia County, Idaho. Simplified from Smith (1982).

and Bruce R. Wardlaw, written commun., 1975–1979, to J. F. Smith, Jr.). Local recrystallization of the Pennsylvanian limestone is also evident.

IGNEOUS ROCKS

Igneous rocks in the area of the Black Pine Mountains comprise a few small, altered intermediate to mafic dikes, remnants of Tertiary ash-fall beds and tuffaceous sedimentary strata, and rhyolitic ash-flow tuffs; lithoidal rhyolites crop out in the topographic expression of a rhyolite dome about 5 km southwest of the mountains. The radiometric age of these glassy intrusive and extrusive rocks in the dome was determined by C. W. Naeser, using the fission-track method on zircons, as 8.3 ± 1.7 m.y. (Williams and others, 1976). Dikes of presumably Tertiary age are poorly exposed in the mining district. Two mafic dikes in an artificial cut in the NE¼ sec. 34, T. 15 S., R. 29 E. do not exceed 1 m in width. They were intruded along faults that strike north and have vertical dips. The dikes are extensively weathered and do not form continuous exposures. Contacts between the dikes and the country rocks are sharp, and no prominent thermal effects were observed in the host sedimentary strata.

The mafic dikes are porphyritic and have an intersertal texture, which is locally obliterated by alteration. The principal constituents of the dikes are plagioclase as phenocrysts and pyroxene and biotite in the groundmass. Secondary minerals or oxidation products include quartz, calcite, muscovite, pyrite, magnetite, hematite, rutile, and trace amounts of chlorite, penninite, and epidote. Plagioclase with the composition of labradorite occurs as twinned subhedral crystals, which are commonly less than 1 mm long. The interior of the plagioclase crystals is commonly completely replaced by quartz, calcite, and minor amounts of sericite, and the margins are also locally embayed by quartz. Subhedral crystals of pyroxene have a glomeroporphyritic texture in places. Biotite is extensively leached of iron, and it is commonly completely altered to muscovite. Silicification preceded calcification in the dikes, and both quartz and calcite occur as microveinlets or irregular replacement masses throughout the intrusives.

GENERAL STRUCTURE

Low-angle faults, along which younger rocks have been moved over older ones or over rocks of partly equivalent ages, form the most prominent structural features in the Black Pine Mountains. They are more prevalent than the steeper normal faults, which may, however, be more abundant than can be shown on the geologic map, as their recognition is hampered by poor exposures in many places and by the lack of traceable key beds. Normal faults seem to have no pre-ferred orientations.

Both large- and small-scale folds appear to be common in the Black Pine Mountains, but delineation of trends is difficult in many parts of the area because bedding attitudes differ in short distances. The folds, whose axial trends are not shown on figure 2, comprise anti-clines and synclines of principally north and northeast trends.

MINERAL DEPOSITS

VEIN DEPOSITS

The vein deposits are characterized by quartz veins that fill frac-tures chiefly in Pennsylvanian carbonate rocks. These veins are com-monly elongated, steeply inclined tabular bodies that exhibit pro-nounced variations in thickness and lateral continuity. Calcite and minor amounts of barite are common gangue minerals associated with the quartz. Isolated pods and stringers of ore contain tetrahedrite (freibergite), sphalerite, jamesonite, minor pyrite, and uncommonly cinnabar. Secondary oxidation products include smithsonite, hemimor-phite, azurite, malachite, scorodite, bindheimite, and various oxides of antimony and iron. These secondary minerals generally are lacking within 23–30 m of the surface, and supergene processes evidently were not important factors in enriching the ore (Anderson, 1931).

The Silver Hills, Hazel Pine, and Ruth mines evidently were the largest producers of vein-deposit ore in the Black Pine mining district (Anderson, 1931, p. 134–139). The Silver Hills mine, which has the most extensive workings, was inaccessible during my investigation in 1976; it is the best example of a typical quartz vein deposit in the district. Production from this mine, which ceased in 1932, probably did not exceed 900 t (metric tons) of ore. This estimate was made by combining analytical data published by Anderson (1931) with pro-duction figures compiled by the Idaho Bureau of Mines (Campbell, 1924, 1925, 1926, 1932). Lead, silver, gold, and copper, in order of decreasing dollar value, constitute the most important metals in the ore from the Silver Hills mine. Small shipments of siliceous silver ore with lesser values of gold, lead, and zinc were made from the Hazel Pine mine between 1915 and 1917 (Gerry, 1917, 1918, 1919, 1920). About 12 t of zinc carbonate ore was shipped from the Ruth mine in 1915 (Gerry, 1917), and a minor, but unknown, amount of smithsonite ore was also produced from other claims before 1921 (Sie-benthal, 1917; and Gerry, 1918, 1919, 1920).

DEPOSITS OF DISSEMINATED GOLD

Fine-grained disseminated gold occurs in altered Pennsylvanian silty carbonate rock. In 1943, the Virmyra Mining Company located 20 claims near the Tolman mine (fig. 2) in the Strevell 15-minute quadrangle. These claims were called the Virginia group (Campbell, 1943). This property was leased and mined by the Duvall Company from 1949 to 1955 (McDowell, 1949, 1950, 1951, 1952, 1953, 1954, 1955). Workings at the mine consisted of a crescent-shaped, generally east trending open cut, which was almost 180 m long and 90 m wide by 35 m deep, and of four short drifts into the north face of the pit. The ore was processed in a small mill erected near the mine in 1950 (McDowell, 1950). No published production data are available. Several companies have conducted drilling programs in the area and have explored an area of several square kilometers since the early 1960's, but no published data on the results are available. Samples on which most of the following discussion is based were collected from the Duvall Company pit and the immediate vicinity.

HOST ROCKS FOR DEPOSITS OF DISSEMINATED GOLD

Pennsylvanian strata of predominantly sandy and calcareous siltstones and silty limestone and claystone in lenses constitute the host for the deposits of disseminated gold. Siltstone is the most common rock type in what appears to be the principal area of altered rock. These siltstones are thinly laminated red-brown to yellow-brown rocks that are poorly sorted and commonly contain a few percent of very fine to fine quartz sand. The siltstones consist primarily of subangular to subrounded grains of quartz with variable amounts of interstitial clay. Small quantities of calcite occur as localized patches of granular cement. Magnetite, hematite, goethite, and leucoxene, the main accessory minerals, range in diameter from 0.01 mm to 2 mm and occur as trains or small clusters distributed along the bedding planes. The siltstones contain numerous subrounded to lenticular microcavities, which were formed by partial to complete dissolution of magnetite during diagenetic alteration of the sediments. These solution cavities, which range from 0.1 mm to 2.5 mm across, in places contain relict grains of magnetite and commonly are filled with chalcedony. In some cavities, both chalcedony and magnetite have been partially replaced by irregular patches of calcite.

Thin lenses or beds of green to gray or black dense limestone, generally less than a few meters in thickness, are interbedded with the clastic sedimentary rocks. These limestones are mostly silt bearing, partly silicified, partly recrystallized, fossiliferous micrites. Clastic components are randomly dispersed in the micrite matrix. Opaque iron oxides have been partly dissolved, but oxidation and liberation of iron were not as pronounced in the limestones as they were in the clastic rocks. Pseudo-spar or recrystallized calcite as well as fine-grained silica developed during the postdepositional alteration of the sediments.

Massive green to greenish-gray claystones are exposed along the lower part of the north side of the Duvall Company open cut. These claystone units are generally less than 7 m thick. Illite and mixed-layer illite-montmorillonite are the principal constituents of the claystones. Magnetite, hematite, and quartz silt occur in small quantities as accessory constituents. Iron oxide forms a conspicuous yellow-brown coating along the numerous microfractures in these beds. The claystones commonly contain elongated microcavities, which have been filled to varying degrees with chalcedony, in turn locally replaced by incipient calcite crystals. Alteration in the claystones is further characterized by the random development of irregular patches of replacement quartz in the rock matrix.

Variable amounts of carbonaceous material occur as minor constituents in the Pennsylvanian siltstones and limestones, which are exposed in the open cut. Two types of carbonaceous material—graphite and organic matter—were recognized in these sediments. Carbon analyses determined from samples of each rock type (table 1) indicate that carbonate rocks contain slightly higher amounts of organic matter than do the noncalcareous clastic rocks, which presumably are more permeable and more susceptible to oxidation.

OCCURRENCE OF THE GOLD

Finely divided gold and pyrite are the principal metallic minerals in the deposits. They seem to be disseminated chiefly in siltstone, which is commonly oxidized to some degree, and are associated with a gangue that includes calcite, barite, and quartz. Most of the gold grains are submicrometer in size, as shown by electron microscope, and are commonly less than 0.5 μm in diameter. They are disseminated within the clay or silt matrix of the clastic rocks or in the micrite groundmass of the limestones. Some gold is associated with the organic matter in both the clastic and carbonate sedimentary rocks; a seam of carbonaceous siltstone, generally less than 1 m thick, contains the highest concentration of gold in the rocks sampled. The highest gold content of samples from this siltstone was 70 ppm (2.25 oz troy/t). In general, however, no positive association between gold content and the amount of organic matter in the sedimentary rocks could be demonstrated. Oxidized siltstones stratigraphically above this carbonaceous zone contain between 5 and 10 ppm (0.16 and 0.32 oz

TABLE 1.—*Analyses showing the relationship between organic carbon and gold content in altered rocks from the Duvall Company open cut (site of Tolman mine on fig. 2), Black Pine Mountains, Cassia County, Idaho*

[Total carbon determined by thermal conductivity by V. E. Shaw; carbonate carbon determined gasometrically and organic carbon determined by difference, by P. H. Briggs; gold determined by atomic absorption, by J. D. Hoffman; <, less than]

Sample No.	Field No. BS–	Total C	Carbonate C	Organic C	Au (ppm)
			(percent)		
1	19	10.09	7.74	2.35	0.05
2	22	.18	.04	.14	1.5
3	26	.12	<.01	.12	9.5
4	32	9.56	9.18	.38	.35
5	34	.12	<.01	.12	.20
6	35	3.89	3.71	.21	6.5

SAMPLE DESCRIPTIONS

1 Black, carbonaceous, argillaceous, silty limestone; abundant thin calcite veinlets. North side of open cut.

2 Brown argillaceous siltstone; few thin calcite veinlets; rare thin quartz veinlets; trace barite. Adit on north side of open cut.

3 Yellow-brown argillaceous siltstone; moderate amounts of iron oxide; rare thin calcite veinlets. Adit on north side of open cut.

4 Dark-gray, brecciated, silty limestone; abundant calcite veinlets; moderate pyrite; minor barite. Adit on north side of open cut.

5 Red-brown argillaceous siltstone; rare calcite veinlets. Adit on north side of open cut.

6 Gray brecciated siltstone; minor amounts of iron oxide; rare pyrite. Adit on north side of open cut.

troy/t) of gold (table 2). Samples of fault gouge and breccia contain as much as 36 ppm gold (1.15 oz troy/t) (table 2, No. 4).

ASSOCIATED SULFIDE MINERALS

Pyrite occurs commonly in the oxidized rocks, but it does not generally constitute more than 1 percent of the host rock. The pyrite is not readily seen in these rocks because of its small grain size, as most grains are less than 10 μm in diameter; a few grains noted were as large as 20 μm.

Tetrahedrite is the only base-metal sulfide that was identified in the gold-bearing strata, but it was not observed to be associated closely with the gold. The tetrahedrite occurs sparingly as minute grains disseminated in calcite veinlets. Green copper stains in close proximity to tetrahedrite occur uncommonly and in limited areas only.

TABLE 2.—Analyses of altered and unaltered rocks from the Duvall Company open cut (site of Tolman mine on fig. 2) and vicinity, Black Pine mining district, Black Pine Mountains, Cassia County, Idaho

[Numbers in parentheses indicate sensitivity limit of method used. G, undetermined amount of the element is present above number shown; L, undetermined amount of the element is present below sensitivity limit; N, element looked for but not found; ND, not determined. Elements were determined by semiquantitative spectroscopy unless otherwise noted. Analysts for samples: semiquantitative spectrographic analyses by Steve Sutley and J. Motooka; gold analyses by J. D. Hoffman and J. Sharkey[1], antimony analyses by C. A. Curtis, J. Sharkey, and J. Vieta[1]; arsenic analyses by J. Sharkey[2]; mercury analyses by J. G. Vieta and J. G. Frisken[3]. All elements shown under sample TM-9 were determined by semiquantitative spectroscopy by Chris Heropoulos. Fe, Mg, Ca, Ti in percent; others in parts per million]

Field No.—	BS-18	BS-19	BS-23	BS-24	BS-32	BS-34	BS-44	BS-51	BS-52	BS-53	BS-54	BS-55	BS-56	TM-9	BS-43	BS-50
Sample No.—	1	2	3	4	5	6	7	8	9	10	11	12	13	14	15	16
Fe (.05)	1.5	0.15	10	2	0.07	2	10	1	7	10	3	3	1.5		0.05	7
Mg (.02)	.2	1	.1	.05	3	.3	1	.1	.5	.5	.3	.5	.5	.2	.7	1
Ca (.05)	5	10	.5	1	G20	.5	10	.5	.5	.2	1.5	.2	7	1	20	15
Ti (.002)	.2	.02	.2	.3	.05	.2	.3	.2	1	1	.3	.5	.2	.5	.01	.2
Mn (10)	20	50	200	L	200	10	500	150	20	20	10	20	300	15	50	100
Ag (.5)	1	.5	10	7	.5	2	5	2	2	5	1.5	2	1.5	10	.7	3
As (1)	1,200	60	800	2,400	100	80	200	3,300	200	1,600	200	3,300	200	7,000	L10	300
Au (.05)	.05	.05	19	36	.35	.2	L.05	1	.2	7	2.5	6	.4	70	L.05	1.5
Hg (.01)	10	.65	G50	G50	10	1.2	.3	40	13	G50	6.5	G50	7	ND	2.5	13
B (10)	100	15	70	50	20	200	100	50	300	300	100	200	70	70	L	70
Ba (20)	150	N	G5,000	G5,000	700	150	150	G5,000	200	200	300	200	G5,000	30,000	70	200
Be (1)	1.5	L	1	1.5	N	1.5	1	L	2	L	N	1	1	1.5	N	L
Bi (10)	N	N	N	N	N	N	N	N	N	N	N	N	N	N	N	N
Cd (20)	N	N	N	N	N	N	N	N	N	N	N	N	N	N	N	N
Co (5)	5	N	7	L	N	L	70	L	L	L	L	L	10	N	N	5
Cr (10)	150	30	500	700	70	150	1,500	150	1,000	700	700	1,000	200	300	L	150
Cu (5)	15	N	100	50	L	50	150	30	100	200	20	70	50	50	L	100
La (20)	50	N	50	50	L	30	70	L	100	70	70	50	30	70	N	70
Mo (5)	10	5	5	N	N	7	L	N	10	L	L	5	N	N	N	L
Nb (20)	N	N	L	L	N	L	L	N	L	L	L	L	N	5	L	L
Ni (5)	100	5	300	150	20	70	1,000	70	200	150	100	150	150	70	5	200
Pb (10)	L	N	30	20	L	15	20	30	20	30	10	10	10	30	30	30
Sb (.01)	40	5	100	5	30	15	10	60	10	N	5	5	5	300	N	25
Sc (5)	10	L	L	L	5	10	15	N	20	10	5	7	L	5	N	5
Sn (10)	N	N	G5,000	G5,000	1,000	N	N	N	N	N	N	N	N	N	N	N
Sr (10)	100	500	70	100	100	100	700	1,000	L	200	500	300	2,000	2,000	2,000	2,000
V (10)	150	15	N	N	N	200	150	50	700	500	500	1,000	150	150	L	100
W (50)	N	N	N	N	N	N	N	N	N	N	N	N	N	N	N	N
Y (10)	30	L	10	10	30	30	20	15	70	15	15	15	N	15	10	20
Zn (200)	1,000	N	300	500	N	700	N	300	1,000	500	300	500	700	500	N	500
Zr (10)	100	N	50	70	20	100	200	70	150	100	70	100	30	200	15	70

[1] Analyses by atomic absorption.
[2] Analyses by colorimetric method.
[3] Analyses by instrumental detector.

SAMPLE DESCRIPTIONS

1   Yellow-brown, slightly sandy siltstone; much iron oxide; minor calcite veinlets. South side of open cut.
2   Black, carbonaceous, argillaceous, silty limestone; abundant thin calcite veinlets. North side of open cut.
3   Gray, vuggy, brecciated siltstone; moderate amounts of iron oxide; minor barite. Adit C. (Of four adits in the northwest and north side of the open cut, C is the third one from the southwest end.)
4   Yellow-brown brecciated siltstone; much iron oxide; fault gouge. Adit C.
5   Dark-gray, brecciated, slightly silty limestone; abundant calcite veinlets; moderate pyrite; minor barite. Adit C.

6   Red-brown argillaceous siltstone. Adit C.
7   Brown, altered mafic dike; moderate number of calcite veinlets; much iron oxide; abundant pyrite. Slightly north and above main open cut.

8   Green, argillaceous siltstone; minor amounts of iron oxide; minor barite. North side of open cut.
9   Red-brown argillaceous siltstone; much iron oxide. North side of open cut.
10 Green-brown argillaceous siltstone; moderate amounts of iron oxide. North side of open cut.

11 Red-brown, sandy, argillaceous siltstone; much iron oxide. North side of open cut.
12 Greenish-gray, brecciated, calcareous siltstone; moderate amounts of iron oxide; abundant calcite veinlets; few quartz veinlets; rare barite veinlets. North side of open cut.
13 Brown-red, sandy, argillaceous siltstone; much iron oxide. North side of open cut.
14 Carbonaceous siltstone. Bottom of west wall, Adit C.
15 Fossiliferous silty limestone; unaltered Pennsylvanian limestone. Altitude 2,048 m (6,720 ft), 0.3 km northwest of open cut.
16 Light-gray, slightly sandy, calcareous siltstone; minor thin calcite veinlets; unaltered siltstone. About 0.2 km south of open cut.

Trace amounts of cinnabar are widely dispersed in the more porous zones in the gold-bearing sediments. This mineral was also observed as very sparse microscopic crystals or anhedral grains along fractures within brecciated barite veinlets.

Although base-metal sulfides and cinnabar are very sparse as minerals in the oxidized sediments, anomalous amounts of mercury, arsenic, and antimony occur in the samples from the open cut (table 2). The four samples containing 6–36 ppm gold also contain more than 50 ppm mercury and from 800 to 3,300 ppm arsenic. The sample (table 2, No. 14) with the greatest gold content, 70 ppm, contains 7,000 ppm arsenic and 300 ppm antimony; the mercury content of this sample was not determined. These elements and tungsten are associated with gold deposits elsewhere in the Basin and Range province and may be considered as geochemical prospecting indicators in the search for mineralized rock containing gold. Tungsten was not detected (<50 ppm) in the samples in table 2. Anomalous amounts of arsenic, tungsten, and mercury are associated with gold at the Getchell mine north of Winnemucca, Nev. (Erickson and others, 1964); and anomalous amounts of arsenic, mercury, antimony, and tungsten occurred with the gold at the Cortez mine south of Carlin, Nev. (Wells and others, 1969).

GANGUE MINERALS

Barite is locally conspicuous in the mineralized rock, although it constitutes only a minor part of the host sediment. Barite occurs in several forms: bladed crystals in cavities or porous zones; veinlets or discontinuous wispy streaks of finely crystalline barite; and irregular patches that replace matrix carbonate, quartz silt, or clay. No positive correlation is apparent between the occurrence of barite and gold in the mineralized rocks, and sediments that contain anomalous concentrations of barium are not necessarily enriched in gold, although the three samples with the highest gold content are among those with the highest concentrations of barium (table 2).

Secondary silica is noticeable but not abundant in the altered and mineralized rocks. This silica occurs sparingly as thin overgrowths on detrital silt-size quartz grains, as discordant microveinlets, as microscopic replacement patches in the rock matrix, and, locally, in barite.

Calcite has been extensively mobilized in the altered rocks. This mineral occurs principally in veins that are mostly less than 1 cm thick; in some places, only a few veinlets exceed 1 mm in thickness. Some of the larger calcite veins along fractures exhibit a pronounced textural banding in places. Calcite was also noted as small replacement masses in the rock matrix and as incipient crystals replacing quartz and barite.

STRUCTURAL CONTROL OF THE ORE DEPOSITS

Structural features have apparently been important factors in the localization of mineralization in the vicinity of the Duvall Company pit. Both low- and high-angle faults cross the mineralized area and the area nearby. The most prominent structures visible on the northwest wall of the Duvall pit include a well-pronounced subhorizontal fault and shear zone along the lower part of the wall, and several apparently normal steep faults. The prominent low-angle fault essentially defines the backs of three adits in the northwest face of the open cut. This fault is characterized by an undulating surface that in general strikes north-northwest and dips about 35° W. Slickensides along the fault surface in places suggest that the last direction of movement was westward, although major movement in the area was probably eastward. To the top of the pit in part of the hanging wall of this fault, blocks and slabs of limestone as much as about 12 m long and from about 1.5 m to 8 m across are arranged chaotically in tan, pink, and red siltstone and claystone. This jumble of blocks evidently was developed by movement on faults.

More or less paired samples collected across the low-angle fault suggest that gold is more concentrated below the fault than above it, even though this suggestion is based on only six samples and the generalization might not hold for the entire foot and hanging walls. Samples 8, 10, and 12 (table 2) from below the fault contained 1, 7, and 6 ppm gold, and corresponding samples 9, 11, and 13 (table 2) from immediately above the fault contained 0.2, 2.5, and 0.4 ppm gold, respectively. Many steep, normal faults transect mineralized rock mostly above the low-angle fault, and most apparently have small displacement. Fault breccia and gouge collected from several locations contained as much as 36 ppm gold (1/5 oz troy/t) (table 2, No. 4).

CONCLUSIONS

Faults, stratigraphic position, and rock type were important factors in localizing the gold deposits in the Black Pine district. Samples analyzed for this study indicate that oxidized siltstones, which are in a silty limestone part of a Pennsylvanian rock sequence, contain greater concentrations of disseminated fine-grained gold than do other rock types or beds of other ages. Anomalous concentrations of arsenic, antimony, and mercury appear to be associated with gold-bearing rocks. This assemblage of elements, as well as tungsten, is found in other disseminated gold deposits in the Basin and Range province.

The disseminated gold was probably deposited from dilute low-temperature solutions. Fluid inclusions in barite suggest that the temperature of formation was between 125°C and 200°C (C. G. Cunningham, oral commun., 1977). The absence of daughter minerals in the liquid inclusions indicates that the salinity of the fluid was low. Low concentrations of strontium and magnesium determined in barite and calcite, respectively, provide additional evidence for low formation temperature of the deposits. The deposits of fine-grained gold probably formed at depth in a hot spring environment. The Raft River geothermal area is about 20 km west of the mining district. Similar geothermal activity may have formerly occurred in the Black Pine district and served as a source for hot water that mobilized the gold and other metals. Although a few mafic dikes crop out in the district, no surface or geophysical evidence exists to indicate that an intrusive body of any size is buried below or near the district, to have served as a source for the mineralizing solutions.

The age of mineralization in the area is unknown. It certainly is younger than the Permian strata and younger than most of the structural deformation. No good age control, however, is established for the time of deformation in the Black Pine Mountains. In the Raft River Mountains to the southwest, metamorphic deformation was still underway 24.9 ± 0.6 m.y. ago, and extensive eastward movement of allochthonous plates occurred after this deformation (Compton and others, 1977). How closely the deformation in the Black Pine Mountains can be related to that in the Raft River Mountains is still uncertain.

REFERENCES CITED

Anderson, A. L., 1931, Geology and mineral resources of eastern Cassia County, Idaho: Idaho Bureau of Mines and Geology Bulletin 14, 169 p.

Bailey, E. H., 1964, in Mineral and water resources of Idaho: Idaho Bureau of Mines and Geology Special Report 1, U.S. Senate Committee on Interior and Insular Affairs, 355 p.

Campbell, Arthur, 1943, 45th Annual report of the mining industry of Idaho for the year 1943: Annual Report of the State Mine Inspector, p. 157.

Campbell, Stewart, 1924, 26th Annual report of the mining industry of Idaho for the year 1924: Annual Report of the State Mine Inspector, p. 96–97, 234.

———1925, 27th Annual report of the mining industry of Idaho for the year 1925: Annual Report of the State Mine Inspector, p. 115–116, 255.

———1926, 28th Annual report of the mining industry of Idaho for the year 1926: Annual Report of the State Mine Inspector, p. 106–107, 254.

———1932, 34th Annual report of the mining industry of Idaho for the year 1932: Annual Report of the State Mine Inspector, p. 118–119, 288.

Christensen, O. D., 1975, Metamorphism of the Manning Canyon and Chainman Formations: Stanford University Ph. D. thesis, 166 p.

Compton, R. R., Todd, V. R., Zartman, R. E., and Naeser, C. W., 1977, Oligocene and Miocene metamorphism, folding, and low-angle faulting in northwestern Utah: Geological Society of America Bulletin, v. 88, p. 1237–1250.

Erickson, R. L., Marranzino, A. P., Oda, Uteana, and Janes, W. W., 1964, Geochemical exploration near the Getchell mine, Humboldt County, Nevada: U.S. Geological Survey Bulletin 1198–A, p. A1–A26.

French, D. E., 1975, Geology and mineralization of the southeastern part of the Black Pine Mountains, Cassia County, Idaho: Utah State University M.S. thesis, 69 p.

Gerry, C. N., 1917, Gold, silver, copper, lead, and zinc in Idaho and Washington for 1915: U.S. Geological Survey, Mineral Resources, 1915, pt. 1, p. 523–575.

——1918, Gold, silver, copper, lead, and zinc in Idaho and Washington for 1916: U.S. Geological Survey, Mineral Resources, 1916, pt. 1, p. 565–616.

——1919, Gold, silver, copper, lead, and zinc in Idaho and Washington for 1917: U.S. Geological Survey, Mineral Resources, 1919, pt. 1, p. 457–507.

——1920, Gold, silver, copper, lead, and zinc in Idaho and Washington for 1918: U.S. Geological Survey, Mineral Resources, 1918, pt.1, p. 461–511.

Hubbard, C. R., 1955, A survey of the mineral resources of Idaho (with map): Idaho Bureau of Mines and Geology Pamphlet 105, 74 p.

Larsen, E. S., 1919, The occurrence of cinnabar near Black Pine, Idaho, in Livingston, D. C., ed., Tungsten, cinnabar, manganese, molybdenum, and tin deposits of Idaho: Idaho University School of Mines Bulletin 2, v. 14, p. 65–67.

McDowell, G. A., 1949, 51st annual report of the mining industry of Idaho for the year 1949: Annual report of the State Mine Inspector, p. 146.

——1950, 52d annual report of the mining industry of Idaho for the year 1950: Annual Report of the State Mine Inspector, p. 148–149.

——1951, 53d annual report of the mining industry of Idaho for the year 1951: Annual Report of the State Mine Inspector, p. 45.

——1952, 54th annual report of the mining industry of Idaho for the year 1952: Annual Report of the State Mine Inspector, p. 85.

——1953, 55th annual report of the mining industry of Idaho for the year 1953: Annual Report of the State Mine Inspector, p. 112.

——1954, 56th annual report of the mining industry of Idaho for the year 1954: Annual Report of the State Mine Inspector, p. 95.

——1955, 57th annual report of the mining industry of Idaho for the year 1955: Annual Report of the State Mine Inspector, p. 93.

McKaskey, H. D., 1917, Quicksilver for 1915: U.S. Geological Survey, Mineral Resources, 1915, pt. 1, p. 271.

Ransome, F. L., 1921, Quicksilver for 1920: U.S. Geological Survey, Mineral Resources, 1920, pt. 1, p. 419–439.

Ross, C. P., 1941, The metal and coal mining districts of Idaho—with notes on the non-metallic mineral resources of the state (Parts I, II, III): Idaho Bureau of Mines and Geology Pamphlet, no. 57, 263 p.

Ross, S. H., and Savage, C. N., 1967, Idaho earth science: Idaho Bureau of Mines and Geology, Earth Science Series 1, 271 p.

Shannon, E. V., 1926, The minerals of Idaho: U.S. National Museum Bulletin 131, 483 p.

Siebenthal, C. E., 1917, Zinc for 1915: U.S. Geological Survey, Mineral Resources, 1915, pt. 1, p. 851–977.

Smith, J. F., Jr., 1982, Geologic map of the Strevell quadrangle, Cassia County, Idaho: U.S. Geological Survey Miscellaneous Investigations Map I–1403.

——1983, Paleozoic rocks in the Black Pine Mountains, Cassia County, Idaho: U.S. Geological Survey Bulletin 1536.

Varley, Thomas, Wright, C. A., Soper, E. K., and Livingston, D. C. 1919, A preliminary report on the mining districts of Idaho: U.S. Bureau of Mines Bulletin 166, p. 86.

Wells, J. D., Stoiser, L. R., and Elliott, J. E., 1969, Geology and geochemistry of the Cortez gold deposit, Nevada: Economic Geology, v. 64, no. 5, p. 526–537.

Williams, P. L., Mabey, D. R., Zohdy, A. A. R., Ackermann, H., Hoover, D. B., Pierce, K. L., and Oriel, S. S., 1976, Geology and geophysics of the southern Raft River Valley geothermal area, Idaho, U.S.A.: Second United Nations Symposium on the development and use of geothermal resources, San Francisco, California, May 20–29, 1975, v. 2, p. 1273–1282.

BLACK PINE MINES

Silver lode discoveries not far from the Kelton Road brought
a modest array of prospectors to Black Pine at a time when the
Snake River fine gold excitement of 1869 accounted for mining
interest along other parts of that important stage and freight
route. Unlike many of Idaho's gold and silver lodes, Black Pine
(an isolated high butte above Raft River) did not suffer as a
location remote from rail or wagon transportation. Technological
capability to handle complex silver ores still had to be
developed at that time, so Black Pine could not profit greatly
from availability of superior transportation. Promotion of an
1870 property there, assisted by additional discoveries the next
year, finally brought Black Pine more attention than that area
deserved. A Kelton correspondent informed the Dailey Corinne
Reporter, September 26, 1871 of local Black Pine excitement then
current:

> "All is excitement here about the Black Pine Mines,
> every speculative and unoccupied man has gone there.
> We have news that an old location, more than a year
> ago, has just been bonded for a large amount, and as I
> write, a citizen of our town has arrived with specimens
> from a new ledge just discovered some three miles away
> from the old locations, which he avers is the biggest
> thing yet, in fact the country is not prospected at
> all; some time ago, during the Snake River excitement,
> Doc. Rice, Mr. Majors of your place, and a few others,
> made locations but all left them, none worked to
> develop or explore further, except Rice, who has clung
> with a tenacity which is now being rewarded; the old
> locators are hurrying back to save their claims from
> being jumped. I may take a deck passage on a cayuse,
> the coming week, and visit the mines, when I can speak
> more by the card."

Upon returning to Kelton, he described the wonders of Black
Pine in a facetious style often employed in mining accounts of
that time:

> In my last I told you that I should probably inspect
> the Black Pine Mines before writing again, and so, one
> fine morning I started for the hills, distant thirty

miles. It would be needless to tell you of the
charming alkali plains we traversed, dotted with the
picturesque sage, behind every shrub of which peeped a
rabbit or skulked a chicken. Suffice it that we
arrived before dark at the camp, which is situated at
the highest point where water can be obtained. Here we
hammocked for the night, putting up at the Hotel de
Shively, kept by "Jim" and his estimable lady. Up at
daylight for a climb to the mines which are found a
mile above us on top of the ridge, but so steep is the
ascent it takes the workmen one hour and a quarter to
get up to their labors, which are being prosecuted on
the Black Pine and Aerial lodes, owned by Lewis Johnson
& Co., said to be an association of English
capitalists. After reaching the top of the mountains,
which was as near the zenith as I ever expect to be
again, we inspected the mines. Work on the Aeriel
being done sufficient to answer their contract with the
original locators, they have concentrated the forces on
the Black Pine, which is a real fissure lode, with a
shaft down forty feet; and at fifty feet it is the
intention to drift both ways and ship ore steadily,
which will be about Christmas. At that time the wagon
road will be completed to the mines. That, like all
other immensely rich deposits, such as White Pine,
Pioche, etc., are only found on the tops of mountains
where nothing but silver can grown. Veni, vidi, viei:
I came, saw, I got. You see, Judge, I have not
forgotten all the Latin that was flog'd into me. Well,
after prospecting around a little while we found the
biggest thing out--a monster ledge composed of
chlorides, bromides, sulphides and all the other ides,
which is richer than pure silver itself; and now,
instead of having two good feet (which I have been
praying for so long) I have two hundred, and am a
millionaire!

After a decade of relative inactivity, Black Pine revived
somewhat in 1881. Two or three years of exploration resulted in
four small shafts (of ten, thirty, fifty, and sixty feet) along
with some other prospect holes and cuts. Assays ranged from $28
to $800 in silver carbonate, and Alexander Toponce--who had a
notable record in Idaho and Montana mineral development--had
interests there. When he published his autobiography, though,
Toponce did not regard his enterprise here as worthy of mention.
 A fair amount of evidence of mining still remains to be seen at
Black Pine, but this district never attained any great importance
in Idaho's mineral history.

GOLD RUSH BOOKS

OREGON, USA

www.GoldMiningBooks.com

Books On Mining

Visit: www.goldminingbooks.com to order your copies or ask your favorite book seller to offer them.

Mining Books by Kerby Jackson

<u>Gold Dust: Stories From Oregon's Mining Years</u> - Oregon mining historian and prospector, Kerby Jackson, brings you a treasure trove of seventeen stories on Southern Oregon's rich history of gold prospecting, the prospectors and their discoveries, and the breathtaking areas they settled in and made homes. **5" X 8", 98 ppgs. Retail Price: $11.99**

<u>The Golden Trail: More Stories From Oregon's Mining Years</u> - In his follow-up to "Gold Dust: Stories of Oregon's Mining Years", this time around, Jackson brings us twelve tales from Oregon's Gold Rush, including the story about the first gold strike on Canyon Creek in Grant County, about the old timers who found gold by the pail full at the Victor Mine near Galice, how Iradel Bray discovered a rich ledge of gold on the Coquille River during the height of the Rogue River War, a tale of two elderly miners on the hunt for a lost mine in the Cascade Mountains, details about the discovery of the famous Armstrong Nugget and others. **5" X 8", 70 ppgs. Retail Price: $10.99**

Oregon Mining Books

<u>Geology and Mineral Resources of Josephine County, Oregon</u> - Unavailable since the 1970's, this important publication was originally compiled by the Oregon Department of Geology and Mineral Industries and includes important details on the economic geology and mineral resources of this important mining area in South Western Oregon. Included are notes on the history, geology and development of important mines, as well as insights into the mining of gold, copper, nickel, limestone, chromium and other minerals found in large quantities in Josephine County, Oregon. **8.5" X 11", 54 ppgs. Retail Price: $9.99**

<u>Mines and Prospects of the Mount Reuben Mining District</u> - Unavailable since 1947, this important publication was originally compiled by geologist Elton Youngberg of the Oregon Department of Geology and Mineral Industries and includes detailed descriptions, histories and the geology of the Mount Reuben Mining District in Josephine County, Oregon. Included are notes on the history, geology, development and assay statistics, as well as underground maps of all the major mines and prospects in the vicinity of this much neglected mining district. **8.5" X 11", 48 ppgs. Retail Price: $9.99**

<u>The Granite Mining District</u> - Notes on the history, geology and development of important mines in the well known Granite Mining District which is located in Grant County, Oregon. Some of the mines discussed include the Ajax, Blue Ribbon, Buffalo, Continental, Cougar-Independence, Magnolia, New York, Standard and the Tillicum. Also included are many rare maps pertaining to the mines in the area. **8.5" X 11", 48 ppgs. Retail Price: $9.99**

<u>Ore Deposits of the Takilma and Waldo Mining Districts of Josephine County, Oregon</u> - The Waldo and Takilma mining districts are most notable for the fact that the earliest large scale mining of placer gold and copper in Oregon took place in these two areas. Included are details about some of the earliest large gold mines in the state such as the Llano de Oro, High Gravel, Cameron, Platerica, Deep Gravel and others, as well as copper mines such as the famous Queen of Bronze mine, the Waldo, Lily and Cowboy mines. This volume also includes six maps and 20 original illustrations. **8.5" X 11", 74 ppgs.** Retail Price: $9.99

<u>Metal Mines of Douglas, Coos and Curry Counties, Oregon</u> - Oregon mining historian Kerby Jackson introduces us to a classic work on Oregon's mining history in this important re-issue of Bulletin 14C Volume 1, otherwise known as the Douglas, Coos & Curry Counties, Oregon Metal Mines Handbook. Unavailable since 1940, this important publication was originally compiled by the Oregon Department of Geology and Mineral Industries includes detailed descriptions, histories and the geology of over 250 metallic mineral mines and prospects in this rugged area of South West Oregon. **8.5" X 11", 158 ppgs.** Retail Price: $19.99

Metal Mines of Jackson County, Oregon - Unavailable since 1943, this important publication was originally compiled by the Oregon Department of Geology and Mineral Industries includes detailed descriptions, histories and the geology of over 450 metallic mineral mines and prospects in Jackson County, Oregon. Included are such famous gold mining areas as Gold Hill, Jacksonville, Sterling and the Upper Applegate. **8.5" X 11", 220 ppgs. Retail Price: $24.99**

Metal Mines of Josephine County, Oregon - Oregon mining historian Kerby Jackson introduces us to a classic work on Oregon's mining history in this important re-issue of Bulletin 14C, otherwise known as the Josephine County, Oregon Metal Mines Handbook. Unavailable since 1952, this important publication was originally compiled by the Oregon Department of Geology and Mineral Industries includes detailed descriptions, histories and the geology of over 500 metallic mineral mines and prospects in Josephine County, Oregon. **8.5" X 11", 250 ppgs. Retail Price: $24.99**

Metal Mines of North East Oregon - Oregon mining historian Kerby Jackson introduces us to a classic work on Oregon's mining history in this important re-issue of Bulletin 14A and 14B, otherwise known as the North East Oregon Metal Mines Handbook. Unavailable since 1941, this important publication was originally compiled by the Oregon Department of Geology and Mineral Industries and includes detailed descriptions, histories and the geology of over 750 metallic mineral mines and prospects in North Eastern Oregon. **8.5" X 11", 310 ppgs. Retail Price: $29.99**

Metal Mines of North West Oregon - Oregon mining historian Kerby Jackson introduces us to a classic work on Oregon's mining history in this important re-issue of Bulletin 14D, otherwise known as the North West Oregon Metal Mines Handbook. Unavailable since 1951, this important publication was originally compiled by the Oregon Department of Geology and Mineral Industries and includes detailed descriptions, histories and the geology of over 250 metallic mineral mines and prospects in North Western Oregon. **8.5" X 11", 182 ppgs. Retail Price: $19.99**

Mines and Prospects of Oregon - Mining historian Kerby Jackson introduces us to a classic mining work by the Oregon Bureau of Mines in this important re-issue of The Handbook of Mines and Prospects of Oregon. Unavailable since 1916, this publication includes important insights into hundreds of gold, silver, copper, coal, limestone and other mines that operated in the State of Oregon around the turn of the 19th Century. Included are not only geological details on early mines throughout Oregon, but also insights into their history, production, locations and in some cases, also included are rare maps of their underground workings. **8.5" X 11", 314 ppgs. Retail Price: $24.99**

Lode Gold of the Klamath Mountains of Northern California and South West Oregon
(See California Mining Books)

Mineral Resources of South West Oregon - Unavailable since 1914, this publication includes important insights into dozens of mines that once operated in South West Oregon, including the famous gold fields of Josephine and Jackson Counties, as well as the Coal Mines of Coos County. Included are not only geological details on early mines throughout South West Oregon, but also insights into their history, production and locations. **8.5" X 11", 154 ppgs. Retail Price: $11.99**

Chromite Mining in The Klamath Mountains of California and Oregon
(See California Mining Books)

Southern Oregon Mineral Wealth - Unavailable since 1904, this rare publication provides a unique snapshot into the mines that were operating in the area at the time. Included are not only geological details on early mines throughout South West Oregon, but also insights into their history, production and locations. Some of the mining areas include Grave Creek, Greenback, Wolf Creek, Jump Off Joe Creek, Granite Hill, Galice, Mount Reuben, Gold Hill, Galls Creek, Kane Creek, Sardine Creek, Birdseye Creek, Evans Creek, Foots Creek, Jacksonville, Ashland, the Applegate River, Waldo, Kerby and the Illinois River, Althouse and Sucker Creek, as well as insights into local copper mining and other topics. **8.5" X 11", 64 ppgs. Retail Price: $8.99**

Geology and Ore Deposits of the Takilma and Waldo Mining Districts - Unavailable since the 1933, this publication was originally compiled by the United States Geological Survey and includes details on gold and copper mining in the Takilma and Waldo Districts of Josephine County, Oregon. The Waldo and Takilma mining districts are most notable for the fact that the earliest large scale mining of placer gold and copper in Oregon took place in these two areas. Included in this report are details about some of the earliest large gold mines in the state such as the Llano de Oro, High Gravel, Cameron, Platerica, Deep Gravel and others, as well as copper mines such as the famous Queen of Bronze mine, the Waldo, Lily and Cowboy mines. In addition to geological examinations, insights are also provided into the production, day to day operations and early histories of these mines, as well as calculations of known mineral reserves in the area. This volume also includes six maps and 20 original illustrations. **8.5" X 11", 74 ppgs. Retail Price: $9.99**

Gold Mines of Oregon - Oregon mining historian Kerby Jackson introduces us to a classic work on Oregon's mining history in this important re-issue of Bulletin 61, otherwise known as "Gold and Silver In Oregon". Unavailable since 1968, this important publication was originally compiled by geologists Howard C. Brooks and Len Ramp of the Oregon Department of Geology and Mineral Industries and includes detailed descriptions, histories and the geology of over 450 gold mines Oregon. Included are notes on the history, geology and gold production statistics of all the major mining areas in Oregon including the Klamath Mountains, the Blue Mountains and the North Cascades. While gold is where you find it, as every miner knows, the path to success is to prospect for gold where it was previously found. **8.5" X 11", 344 ppgs. Retail Price: $24.99**

Mines and Mineral Resources of Curry County Oregon - Originally published in 1916, this important publication on Oregon Mining has not been available for nearly a century. Included are rare insights into the history, production and locations of dozens of gold mines in Curry County, Oregon, as well as detailed information on important Oregon mining districts in that area such as those at Agness, Bald Face Creek, Mule Creek, Boulder Creek, China Diggings, Collier Creek, Elk River, Gold Beach, Rock Creek, Sixes River and elsewhere. Particular attention is especially paid to the famous beach gold deposits of this portion of the Oregon Coast. **8.5" X 11", 140 ppgs. Retail Price: $11.99**

Chromite Mining in South West Oregon - Originally published in 1961, this important publication on Oregon Mining has not been available for nearly a century. Included are rare insights into the history, production and locations of nearly 300 chromite mines in South Western Oregon. **8.5" X 11", 184 ppgs. Retail Price: $14.99**

Mineral Resources of Douglas County Oregon - Originally published in 1972, this important publication on Oregon Mining has not been available for nearly forty years. Included are rare insights into the geology, history, production and locations of numerous gold mines and other mining properties in Douglas County, Oregon. **8.5" X 11", 124 ppgs. Retail Price: $11.99**

Mineral Resources of Coos County Oregon - Originally published in 1972, this important publication on Oregon Mining has not been available for nearly forty years. Included are rare insights into the geology, history, production and locations of numerous gold mines and other mining properties in Coos County, Oregon. **8.5" X 11", 100 ppgs. Retail Price: $11.99**

Mineral Resources of Lane County Oregon - Originally published in 1938, this important publication on Oregon Mining has not been available for nearly seventy five years. Included are extremely rare insights into the geology and mines of Lane County, Oregon, in particular in the Bohemia, Blue River, Oakridge, Black Butte and Winberry Mining Districts. **8.5" X 11", 82 ppgs. Retail Price: $9.99**

Mineral Resources of the Upper Chetco River of Oregon: Including the Kalmiopsis Wilderness - Originally published in 1975, this important publication on Oregon Mining has not been available for nearly forty years. Withdrawn under the 1872 Mining Act since 1984, real insight into the minerals resources and mines of the Upper Chetco River has long been unavailable due to the remoteness of the area. Despite this, the decades of battle between property owners and environmental extremists over the last private mining inholding in the area has continued to pique the interest of those interested in mining and other forms of natural resource use. Gold mining began in the area in the 1850's and has a rich history in this geographic area, even if the facts surrounding it are little known. Included are twenty two rare photographs, as well as insights into the Becca and Morning Mine, the Emmly Mine (also known as Emily Camp), the Frazier Mine, the Golden Dream or Higgins Mine, Hustis Mine, Peck Mine and others. **8.5" X 11", 64 ppgs. Retail Price: $8.99**

Gold Dredging in Oregon - Originally published in 1939, this important publication on Oregon Mining has not been available for nearly seventy five years. Included are extremely rare insights into the history and day to day operations of the dragline and bucketline gold dredges that once worked the placer gold fields of South West and North East Oregon in decades gone by. Also included are details into the areas that were worked by gold dredges in Josephine, Jackson, Baker and Grant counties, as well as the economic factors that impacted this mining method. This volume also offers a unique look into the values of river bottom land in relation to both farming and mining, in how farm lands were mined, re-soiled and reclamated after the dredges worked them. Featured are hard to find maps of the gold dredge fields, as well as rare photographs from a bygone era. **8.5" X 11", 86 ppgs. Retail Price: $8.99**

Quick Silver Mining in Oregon - Originally published in 1963, this important publication on Oregon Mining has not been available for over fifty years. This publication includes details into the history and production of Elemental Mercury or Quicksilver in the State of Oregon. **8.5" X 11", 238 ppgs. Retail Price: $15.99**

Mines of the Greenhorn Mining District of Grant County Oregon - Originally published in 1948, this important publication on Oregon Mining has not been available for over sixty five years. In this publication are rare insights into the mines of the famous Greenhorn Mining District of Grant County, Oregon, especially the famous Morning Mine. Also included are details on the Tempest, Tiger, Bi-Metallic, Windsor, Psyche, Big Johnny, Snow Creek, Banzette and Paramount Mines, as well as prospects in the vicinities in the famous mining areas of Mormon Basin, Vinegar Basin and Desolation Creek. Included are hard to find mine maps and dozens of rare photographs from the bygone era of Grant County's rich mining history. **8.5" X 11", 72 ppgs. Retail Price: $9.99**

Geology of the Wallowa Mountains of Oregon: Part I (Volume 1) - Originally published in 1938, this important publication on Oregon Mining has not been available for nearly seventy five years. Included are details on the geology of this unique portion of North Eastern Oregon. This is the first part of a two book series on the area. Accompanying the text are rare photographs and historic maps.**8.5" X 11", 92 ppgs. Retail Price: $9.99**

Geology of the Wallowa Mountains of Oregon: Part II (Volume 2) - Originally published in 1938, this important publication on Oregon Mining has not been available for nearly seventy five years. Included are details on the geology of this unique portion of North Eastern Oregon. This is the first part of a two book series on the area. Accompanying the text are rare photographs and historic maps.**8.5" X 11", 94 ppgs. Retail Price: $9.99**

Field Identification of Minerals For Oregon Prospectors - Originally published in 1940, this important publication on Oregon Mining has not been available for nearly seventy five years. Included in this volume is an easy system for testing and identifying a wide range of minerals that might be found by prospectors, geologists and rockhounds in the State of Oregon, as well as in other locales. Topics include how to put together your own field testing kit and how to conduct rudimentary tests in the field. This volume is written in a clear and concise way to make it useful even for beginners. **8.5" X 11", 158 ppgs. Retail Price: $14.99**

The Bohemia Mining District of Oregon - Originally published in 1900, this important publication on Oregon Mining has not been available for over a century. Included in this volume are important insights into the famous Bohemia Mining District of Oregon, including the histories and locations of important gold mines in the area such as the Ophir Mine, Clarence, Acturas, Peek-a-boo, White Swan, Combination Mine, the Musick Mine, The California, White Ghost, The Mystery, Wall Street, Vesuvius, Story, Lizzie Bullock, Delta, Elsie Dora, Golden Slipper, Broadway, Champion Mine, Knott, Noonday, Helena, White Wings, Riverside and others. Also included are notes on the nearby Blue River Mining District. **8.5" X 11", 58 ppgs. Retail Price: $9.99**

The Gold Fields of Eastern Oregon - Unavailable since 1900, this publication was originally compiled by the Baker City Chamber of Commerce Offering important insights into the gold mining history of Eastern Oregon, "The Gold Fields of Eastern Oregon" sheds a rare light on many of the gold mines that were operating at the turn of the 19th Century in Baker County and Grant County in North Eastern Oregon. Some of the areas featured include the Cable Cove District, Baisely-Elhorn, Granite, Red Boy, Bonanza, Susanville, Sparta, Virtue, Vaughn, Sumpter, Burnt River, Rye Valley and other mining districts. Included is basic information on not only many gold mines that are well known to those interested in Eastern Oregon mining history, but also many mines and prospects which have been mostly lost to the passage of time. Accompanying are numerous rare photos **8.5" X 11", 78 ppgs. Retail Price: $10.99**

Gold Mining in Eastern Oregon - Originally published in 1938, this important publication on Oregon Mining has not been available for over a century. Included in this volume are important insights into the famous mining districts of Eastern Oregon during the late 1930's. Particular attention is given to those gold mines with milling and concentrating facilities in the Greenhorn, Red Boy, Alamo, Bonanza, Granite, Cable Cove, Cracker Creek, Virtue, Keating, Medical Springs, Sanger, Sparta, Chicken Creek, Mormon Basin, Connor Creek, Cornucopia and the Bull Run Mining Districts. Some of the mines featured include the Ben Harrison, North Pole-Columbia, Highland Maxwell, Baisley-Elkhorn, White Swan, Balm Creek, Twin Baby, Gem of Sparta, New Deal, Gleason, Gifford-Johnson, Cornucopia, Record, Bull Run, Orion and others. Of particular interest are the mill flow sheets and descriptions of milling operations of these mines. **8.5" X 11", 68 ppgs. Retail Price: $8.99**

The Gold Belt of the Blue Mountains of Oregon - Originally published in 1901, this important publication on Oregon Mining has not been available for over a century. Included in this volume are rare insights into the gold deposits of the Blue Mountains of North East Oregon, including the history of their early discovery and early production. Extensive details are offered on this important mining area's mineralogy and economic geology, as well as insights into nearby gold placers, silver deposits and copper deposits. Featured are the Elkhorn and Rock Creek mining districts, the Pocahontas district, Auburn and Minersville districts, Sumpter and Cracker Creek, Cable Cove, the Camp Carson district, Granite, Alamo, Greenhorn, Robinsonville, the Upper Burnt River Valley and Bonanza districts, Susanville, Quartzburg, Canyon Creek, Virtue, the Copper Butte district, the North Powder River, Sparta, Eagle Creek, Cornucopia, Pine Creek, Lower Powder River, the Upper Snake River Canyon, Rye Valley, Lower Burnt River Valley, Mormon Basin, the Malheur and Clarks Creek districts, Sutton Creek and others. Of particular interest are important details on numerous gold mines and prospects in these mining districts, including their locations, histories, geology and other important information, as well as information on silver, copper and fire opal deposits. **8.5" X 11", 250 ppgs. Retail Price: $24.99**

Mining in the Cascades Range of Oregon - Originally published in 1938, this important publication on Oregon Mining has not been available for over seventy five years. Included in this volume are rare insights into the gold mines and other types of metal mines in the Cascades Mountain Range of Oregon. Some of the important mining areas covered include the famous Bohemia Mining District, the North Santiam Mining District, Quartzville Mining District, Blue River Mining District, Fall Creek Mining District, Oakridge District, Zinc District, Buzzard-Al Sarena District, Grand Cove, Climax District and Barron Mining District. Of particular interest are important details on over 100 mines and prospects in these mining districts, including their locations, histories, geology and other important information. 8.5" X 11", 170 ppgs. Retail Price: $14.99

Beach Gold Placers of the Oregon Coast - Originally published in 1934, this important publication on Oregon Mining has not been available for over 80 years. Included in this volume are rare insights into the beach gold deposits of the State of Oregon, including their locations, occurance, composition and geology. Of particular interest is information on placer platinum in Oregon's rich beach deposits. Also included are the locations and other information on some famous Oregon beach mines, including the Pioneer, Eagle, Chickamin, Iowa and beach placer mines north of the mouth of the Rogue River. 8.5" X 11", 60 ppgs. Retail Price: $8.99

Mineralogical Composition of the Sands of the Oregon Coast: From Coos Bay to the Columbia - Published in 1945, he text features hard to find information on the composition of the gold bearing black sands of the South West Oregon Coast, offering a unique insight to prospectors in search of Oregon's legendary beach gold. 104 ppgs, $9.99

Manganese Mining in Oregon - First released in 1942 and now out of print, this special reprint edition of "Manganese in Oregon" was originally published by the Oregon Department of Geology and Mineral Industries. The text features hard to find information on the mining of Manganese in Oregon, including details and maps of Oregon manganese mines and prospects. 108 ppgs, 9.99

Medford Oregon As A Mining Center - Written in 1912, this hard to find publication includes valuable insights into the mining history of South West Oregon. This small book contains interesting information on the gold, copper and mining industry in Southern Oregon as it existed just prior to World War One, shedding light on some of the important mines in the area. Included are rare photographs and vintage advertising of the day. 80 ppgs, 9.99

Mineral Resources of Curry County Oregon - First released in 1977 and now out of print, this special reprint edition of "Geology, Mineral Resources and Rock Materials of Curry County, Oregon" was originally published in cooperation of Curry County, Oregon and the Oregon Department of Geology and Mineral Industries. The text features hard to find information on not only the mining of gold and other metals in Curry County, but also aggregate mining in the area. 102 ppgs, 11.99

Origin of the Gold Bearing Black Sands of the Coast of South West Oregon - First released in 1943 and now out of print, this special reprint edition of "The Origin of the Black Sands of the South West Oregon Coast" was originally published by the Oregon Department of Geology and Mineral Industries. The text features hard to find information on the origin of the gold bearing black sands of the South West Oregon Coast, offering a unique insight to prospectors in search of Oregon's legendary beach gold. 52 ppgs, 8.99

South West Oregon Mining - Leading mining historian Kerby Jackson introduces us to six classic small mining publications on the Gold Mining Industry in Southern Oregon. This small book consists of a compilation of USGS J.S. Diller's "Mines of the Riddles Quadrangle", "The Rogue River Valley Coal Fields" and "Mineral Resources of the Grants Pass Quadrangle", the Grants Pass Commercial Club's rare publication "Mining in Josephine County, Oregon" and the USGS publication "The Distribution of Placer Gold in the Sixes River, South West Oregon". Also included is F.W. Libbey's legendary article on the Southern Oregon Mining Industry, "Lest We Forget", which appeared in the publication of the Oregon State Department of Geology and Mineral Industries in the early 1960's. This compilation offers a unique perspective on mining in South West Oregon and includes considerable information on mines in Josephine, Jackson and Coos Counties. 142 ppgs, 14.99

Geology and Mineral Resources of the Gasquet Quadrangle of California-Oregon - First published in 1953, it has been unavailable for over a century and sheds important light on the geological features and mineral resources of this portion of Northern California and Southern Oregon. 80 ppgs, 9.99

Idaho Mining Books

Gold in Idaho - Unavailable since the 1940's, this publication was originally compiled by the Idaho Bureau of Mines and includes details on gold mining in Idaho. Included is not only raw data on gold production in Idaho, but also valuable insight into where gold may be found in Idaho, as well as practical information on the gold bearing rocks and other geological features that will assist those looking for placer and lode gold in the State of Idaho. This volume also includes thirteen gold maps that greatly enhance the practical usability of the information contained in this small book detailing where to find gold in Idaho. **8.5" X 11", 72 ppgs. Retail Price: $9.99**

Geology of the Couer D'Alene Mining District of Idaho - Unavailable since 1961, this publication was originally compiled by the Idaho Bureau of Mines and Geology and includes details on the mining of gold, silver and other minerals in the famous Coeur D'Alene Mining District in Northern Idaho. Included are details on the early history of the Coeur D'Alene Mining District, local tectonic settings, ore deposit features, information on the mineral belts of the Osburn Fault, as well as detailed information on the famous Bunker Hill Mine, the Dayrock Mine, Galena Mine, Lucky Friday Mine and the infamous Sunshine Mine. This volume also includes sixteen hard to find maps. **8.5" X 11", 70 ppgs. Retail Price: $9.99**

The Gold Camps and Silver Cities of Idaho - Originally published in 1963, this important publication on Idaho Mining has not been available for nearly fifty years. Included are rare insights into the history of Idaho's Gold Rush, as well as the mad craze for silver in the Idaho Panhandle. Documented in fine detail are the early mining excitements at Boise Basin, at South Boise, in the Owyhees, at Deadwood, Long Valley, Stanley Basin and Robinson Bar, at Atlanta, on the famous Boise River, Volcano, Little Smokey, Banner, Boise Ridge, Hailey, Leesburg, Lemhi, Pearl, at South Mountain, Shoup and Ulysses, Yellow Jacket and Loon Creek. The story follows with the appearance of Chinese miners at the new mining camps on the Snake River, Black Pine, Yankee Fork, Bay Horse, Clayton, Heath, Seven Devils, Gibbonsville, Vienna and Sawtooth City. Also included are special sections on the Idaho Lead and Silver mines of the late 1800's, as well as the mining discoveries of the early 1900's that paved the way for Idaho's modern mining and mineral industry. Lavishly illustrated with rare historic photos, this volume provides a one of a kind documentary into Idaho's mining history that is sure to be enjoyed by not only modern miners and prospectors who still scour the hills in search of nature's treasures, but also those enjoy history and tromping through overgrown ghost towns and long abandoned mining camps. **8.5" X 11", 186 ppgs. Retail Price: $14.99**

Ore Deposits and Mining in North Western Custer County Idaho - Unavailable since 1913, this important publication was originally published by the Us Department of the Interior and has been unavailable for a century. Included are fine details on the geology, geography, gold placers and gold and silver bearing quartz veins of the mining region of North West Custer County, Idaho. Of particular interest is a rare look at the mines and prospects of the region, including those such as the Ramshorn Mine, SkyLark, Riverview, Excelsior, Beardsley, Pacific, Hoosier, Silver Brick, Forest Rose and dozens of others in the Bay Horse Mining District. Also covered are the mines of the Yankee Fork District such as the Lucky Boy, Badger, Black, Enterprise, Charles Dickens, Morrison, Golden Sunbeam, Montana, Golden Gate and others, as well as those in the Loon Mining District. **8.5" X 11", 126 ppgs. Retail Price: $12.99**

Gold Rush To Idaho - Unavailable since 1963, this important publication was originally published by the Idaho Bureau of Mines and has been unavailable for 50 years. "Gold Rush To Idaho" revisits the earliest years of the discovery of gold in Idaho Territory and introduces us to the conditions that the pioneer gold seekers met when they blazed a trail through the wilderness of Idaho's mountains and discovered the precious yellow metal at Oro Fino and Pierce. Subsequent rushes followed at places like Elk City, Newsome, Clearwater Station, Florence, Warrens and elsewhere. Of particular interest is a rare look at the hardships that the first miners in Idaho met with during their day to day existences and their attempts to bring law and order to their mining camps. **8.5" X 11", 88 ppgs. Retail Price: $9.99**

The Geology and Mines of Northern Idaho and North Western Montana - Unavailable since 1909, this important publication was originally published by the Us Department of the Interior and has been unavailable for a century. Included are fine details on the geology and geography of the mining regions of Northern Idaho and North Western Montana. Of particular interest is a rare look at the mines and prospects of the region, including those in the Pine Creek Mining District, Lake Pend Oreille district, Troy Mining District, Sylvanite District, Cabinet Mining District, Prospect Mining District and the Missoula Valley. Some of the mines featured include the Iron Mountain, Silver Butte, Snowshoe, Grouse Mountain Mine and others. **8.5" X 11", 142 ppgs. Retail Price: $12.99**

Mining in the Alturas Quadrangle of Blaine County Idaho - Unavailable since 1922, this important publication was originally published by the Idaho Bureau of Mines and has been unavailable for ninety years. Topics include the geology, rock formations and the formation of ore deposits in this important mining area of Idaho. Of particular focus is information on the local geology, quartz veins and ore deposits of this portion of Idaho. Included are hard to find details, including the descriptions and locations of numerous gold and silver mines in the area including the Silver King, Pilgrim, Columbia, Lone Jack, Sunbeam, Pride of the West, Lucky Boy, Scotia, Atlanta, Beaver-Bidwell and others mines and prospects. **8.5" X 11", 56 ppgs. Retail Price: $8.99**

Mining in Lemhi County Idaho - Originally published in 1913, this important book on Idaho Mining has not been available to miners for over a century. Included are rare insights into hundreds of gold, silver, copper and other mines in this famous Idaho mining area. Details include the locations, geology, history, production and other facts of the mines of this region, not only gold and silver hardrock mines, but also gold placer mines, lead-silver deposits, copper mines, cobalt-nickel deposits, tungsten and tin mines . It is lavishly illustrated with hard to find photos of the period and rare mining maps. Some of the vicinities featured include the Nicholia Mining District, Spring Mountain District, Texas District, Blue Wing District, Junction District, McDevitt District, Pratt Creek, Eldorado District, Kirtley Creek, Carmen Creek, Gibbonsville, Indian Creek, Mineral Hill District, Mackinaw, Eureka District, Blackbird District, YellowJacket District, Gravel Range District, Junction District, Parker Mountain and other mining districts. **8.5″ X 11″, 226 ppgs. Retail Price: $19.99**

Mining in Shoshone County Idaho - First published in 1923, it has been unavailable for over a century and sheds important light on the mining history of Shoshone County, Idaho. Some of the topics include the history of mining in Shoshone County, a look at the local geology and ore characteristics of lead-silver deposits, zinc deposits, copper, antimony, gold and other minerals. Also included are insights into the history, production, characteristics and locations of numerous mines in the area. 198 ppgs, 15.99

Utah Mining Books

Fluorite in Utah - Unavailable since 1954, this publication was originally compiled by the USGS, State of Utah and U.S. Atomic Energy Commission and details the mining of fluorspar, also known as fluorite in the State of Utah. Included are details on the geology and history of fluorspar (fluorite) mining in Utah, including details on where this unique gem mineral may be found in the State of Utah. 8.5″ X 11″, 60 ppgs. Retail Price: $8.99

The Gold Hill Mining District of Utah - First published in 1935, it has been unavailable since those days and sheds important light on the mines, history and geology of Utah's Gold Hill Mining District. Included are rare insights into this important mining area, including the locations, histories and details of numerous mines. This volume is well illustrated with geological diagrams, as well as hard to find maps of some of the most important mines in this district. 202 ppgs., 19.99

The Mines, Miners and Minerals of Utah - First published in 1896, it has been unavailable since those days and sheds important light on the early mines and miners of Pioneer Utah, as well as the minerals which they won from the earth by laborious hard physical labor and sheer determination. Included are rare insights into the early mining history of Utah, as well details on hundreds of gold, silver and copper mines. 376 ppgs., 24.99

California Mining Books

The Tertiary Gravels of the Sierra Nevada of California - Mining historian Kerby Jackson introduces us to a classic mining work by Waldemar Lindgren in this important re-issue of The Tertiary Gravels of the Sierra Nevada of California. Unavailable since 1911, this publication includes details on the gold bearing ancient river channels of the famous Sierra Nevada region of California. 8.5″ X 11″, 282 ppgs. Retail Price: $19.99

The Mother Lode Mining Region of California - Unavailable since 1900, this publication includes details on the gold mines of California's famous Mother Lode gold mining area. Included are details on the geology, history and important gold mines of the region, as well as insights into historic mining methods, mine timbering, mining machinery, mining bell signals and other details on how these mines operated. Also included are insights into the gold mines of the California Mother Lode that were in operation during the first sixty years of California's mining history. 8.5″ X 11″, 176 ppgs. Retail Price: $14.99

Lode Gold of the Klamath Mountains of Northern California and South West Oregon - Unavailable since 1971, this publication was originally compiled by Preston E. Hotz and includes details on the lode mining districts of Oregon and California's Klamath Mountains. Included are details on the geology, history and important lode mines of the French Gulch, Deadwood, Whiskeytown, Shasta, Redding, Muletown, South Fork, Old Diggings, Dog Creek (Delta), Bully Choop (Indian Creek), Harrison Gulch, Hayfork, Minersville, Trinity Center, Canyon Creek, East Fork, New River, Denny, Liberty (Black Bear), Cecilville, Callahan, Yreka, Fort Jones and Happy Camp mining districts in California, as well as the Ashland, Rogue River, Applegate, Illinois River, Takilma, Greenback, Galice, Silver Peak, Myrtle Creek and Mule Creek districts of South Western Oregon. Also included are insights into the mineralization and other characteristics of this important mining region. 8.5″ X 11″, 100 ppgs. Retail Price: $10.99

Mines and Mineral Resources of Shasta County, Siskiyou County, Trinity County: California - Unavailable since 1915, this publication was originally compiled by the California State Mining Bureau and includes details on the gold mines of this area of Northern California. Also included are insights into the mineralization and other characteristics of this important mining region, as well as the location of historic gold mines. **8.5″ X 11″, 204 ppgs. Retail Price: $19.99**

Geology of the Yreka Quadrangle, Siskiyou County, California - Unavailable since 1977, this publication was originally compiled by Preston E. Hotz and includes details on the geology of the Yreka Quadrangle of Siskiyou County, California. Also included are insights into the mineralization and other characteristics of this important mining region. **8.5" X 11", 78 ppgs. Retail Price: $7.99**

Mines of San Diego and Imperial Counties, California - Originally published in 1914, this important publication on California Mining has not been available for a century. This publication includes important information on the early gold mines of San Diego and Imperial County, which were some of the first gold fields mined in California by early Spanish and Mexican miners before the 49ers came on the scene. Included are not only details on early mining methods in the area, production statistics and geological information, but also the location of the early gold mines that helped make California "The Golden State". Also included are details on the mining of other minerals such as silver, lead, zinc, manganese, tungsten, vanadium, asbestos, barite, borax, cement, clay, dolomite, fluospar, gem stones, graphite, marble, salines, petroleum, stronium, talc and others. **8.5" X 11", 116 ppgs. Retail Price: $12.99**

Mines of Sierra County, California - Unavailable since 1920, this publication was originally compiled by the California State Mining Bureau and includes details on the gold mines of Sierra County, California. Also included are insights into the mineralization and other characteristics of this important mining region, as well as the location of historic gold mines. **8.5" X 11", 156 ppgs. Retail Price: $19.99**

Mines of Plumas County, California - Unavailable since 1918, this publication was originally compiled by the California State Mining Bureau and includes details on the gold mines of Plumas County, California. Also included are insights into the mineralization and other characteristics of this important mining region, as well as the location of historic gold mines. **8.5" X 11", 200 ppgs. Retail Price: $19.99**

Mines of El Dorado, Placer, Sacramento and Yuba Counties, California - Originally published in 1917, this important publication on California Mining has not been available for nearly a century. This publication includes important information on the early gold mines of El Dorado County, Placer County, Sacramento County and Yuba County, which were some of the first gold fields mined by the Forty-Niners during the California Gold Rush. Included are not only details on early mining methods in the area, production statistics and geological information, but also the location of the early gold mines that helped make California "The Golden State". Also included are insights into the early mining of chrome, copper and other minerals in this important mining area. **8.5" X 11", 204 ppgs. Retail Price: $19.99**

Mines of Los Angeles, Orange and Riverside Counties, California - Originally published in 1917, this important publication on California Mining has not been available for nearly a century. This publication includes important information on the early gold mines of Los Angeles County, Orange County and Riverside County, which were some of the first gold fields mined in California by early Spanish and Mexican miners before the 49ers came on the scene. Included are not only details on early mining methods in the area, production statistics and geological information, but also the location of the early gold mines that helped make California "The Golden State". **8.5" X 11", 146 ppgs. Retail Price: $12.99**

Mines of San Bernadino and Tulare Counties, California - Originally published in 1917, this important publication on California Mining has not been available for nearly a century. This publication includes important information on the early gold mines of San Bernadino and Tulare County, which were some of the first gold fields mined in California by early Spanish and Mexican miners before the 49ers came on the scene. Included are not only details on early mining methods in the area, production statistics and geological information, but also the location of the early gold mines that helped make California "The Golden State". Also included are details on the mining of other minerals such as copper, iron, lead, zinc, manganese, tungsten, vanadium, asbestos, barite, borax, cement, clay, dolomite, fluospar, gem stones, graphite, marble, salines, petroleum, stronium, talc and others. **8.5" X 11", 200 ppgs. Retail Price: $19.99**

Chromite Mining in The Klamath Mountains of California and Oregon - Unavailable since 1919, this publication was originally compiled by J.S. Diller of the United States Department of Geological Survey and includes details on the chromite mines of this area of Northern California and Southern Oregon. Also included are insights into the mineralization and other characteristics of this important mining region, as well as the location of historic mines. Also included are insights into chromite mining in Eastern Oregon and Montana. **8.5" X 11", 98 ppgs. Retail Price: $9.99**

Mines and Mining in Amador, Calaveras and Tuolumne Counties, California - Unavailable since 1915, this publication was originally compiled by William Tucker and includes details on the mines and mineral resources of this important California mining area. Included are details on the geology, history and important gold mines of the region, as well as insights into other local mineral resources such as asbestos, clay, copper, talc, limestone and others. Also included are insights into the mineralization and other characteristics of this important portion of California's Mother Lode mining region. **8.5" X 11", 198 ppgs. Retail Price: $14.99**

The Cerro Gordo Mining District of Inyo County California - Unavailable since 1963, this publication was originally compiled by the United States Department of Interior. Included are insights into the mineralization and other characteristics of this important mining region of Southern California. Topics include the mining of gold and silver in this important mining district in Inyo County, California, including details on the history, production and locations of the Cerro Gordo Mine, the Morning Star Mine, Estelle Tunnel, Charles Lease Tunnel, Ignacio, Hart, Crosscut Tunnel, Sunset, Upper Newtown, Newtown, Ella, Perseverance, Newsboy, Belmont and other silver and gold mines in the Cerro Gordo Mining District. This volume also includes important insights into the fossil record, geologic formations, faults and other aspects of economic geology in this California mining district. **8.5" X 11", 104 ppgs. Retail Price: $10.99**

Mining in Butte, Lassen, Modoc, Sutter and Tehama Counties of California - Unavailable since 1917, this publication was originally compiled by the United States Department of Interior. Included are insights into the mineralization and other characteristics of this important mining region of California. Topics include the mining of asbestos, chromite, gold, diamonds and manganese in Butte County, the mining of gold and copper in the Hayden Hill and Diamond Mountain mining districts of Lassen County, the mining of coal, salt, copper and gold in the High Grade and Winters mining districts of Modoc County, gold mining in Sutter County and the mining of gold, chromite, manganese and copper in Tehama County. This volume also includes the production records and locations of numerous mines in this important mining region. **8.5" X 11", 114 ppgs. Retail Price: $11.99**

Mines of Trinity County California - Originally published in 1965, this important publication on California Mining has not been available for nearly fifty years. This publication includes important information on mines and mining in Trinity County, California, as well insights into the mineralization and geology of this important mining area in Northern California. Included are extensive details on hardrock and placer gold mines and prospects, including charts showing the locations of these historic mines.. **8.5" X 11", 144 ppgs. Retail Price: $12.99**

Mines of Kern County California - Originally published in 1962, this important publication on California Mining has not been available for nearly fifty years. This publication includes important information on mines and mining in Kern County, California, as well insights into the mineralization and geology of this important mining area in California. Included are extensive details on hardrock and placer gold mines and prospects, including charts showing the locations of these historic mines. **8.5" X 11", 398 ppgs. Retail Price: $24.99**

Mines of Calaveras County California - Originally published in 1962, this important publication on California Mining has not been available for nearly fifty years. This publication includes important information on mines and mining in Calaveras County, California, as well insights into the mineralization and geology of this important mining area in Northern California. Included are extensive details on hardrock and placer gold mines and prospects, including charts showing the locations of these historic mines. **8.5" X 11", 236 ppgs. Retail Price: $19.99**

Lode Gold Mining in Grass Valley California - Unavailable since 1940, this publication was originally compiled by the United States Department of Interior. Included are insights into the gold mineralization and other characteristics of this important mining region of Nevada County, California. This volume also includes important insights into the geologic formations, faults and other aspects of economic geology in this California mining district. Of particular interest are the fine details on many hardrock gold mines in the area, including their locations, histories, development and mineralization. Some of the mines featured include the Gold Hill Mine, Massachusetts Hill, Boundary, Peabody, Golden Center, North Star, Omaha, Lone Jack, Homeward Bound, Hartery, Wisconsin, Allison Ranch, Phoenix, Kate Hayes, W.Y.O.D., Empire, Rich Hill, Daisy Hill, Orleans, Sultana, Centennial, Conlin, Ben Franklin, Crown Point and many others. **8.5" X 11", 148 ppgs. Retail Price: $12.99**

Lode Mining in the Alleghany District of Sierra County California - Unavailable since 1913, this publication was originally compiled by the United States Department of Interior. Included are insights into the mineralization and other characteristics of this important mining region of Sierra County. Included are details on the history, production and locations of numerous hardrock gold mines in this famous California area, including the Tightner Mine, Minnie D., Osceola, Eldorado, Twenty One, Sherman, Kenton, Oriental, Rainbow, Plumbago, Irelan, Gold Canyon, North Fork, Federal, Kate Hardy and others. This volume also includes important insights into the fossil record, geologic formations, faults and other aspects of economic geology in this California mining district. **8.5" X 11", 48 ppgs. Retail Price: $7.99**

Six Months In The Gold Mines During The California Gold Rush - Unavailable since 1850, this important work is a first hand account of one "49'ers" personal experience during the great California Gold Rush, shedding important light on one of the most exciting periods in the history of not only California, but also the world. Compiled from journals written between 1847 and 1849 by E. Gould Buffum, a native of New York, "Six Months In The Gold Mines During The California Gold Rush" offers a rare look into the day to day lives of the people who came to California to work in her gold mines when the state was still a great frontier. **8.5" X 11", 290 ppgs. Retail Price: $19.99**

<u>**Quartz Mines of the Grass Valley Mining District of California**</u> - Unavailable since 1867, this important publication has not been available since those days. This rare publication offers a short dissertation on the early hardrock mines in this important mining district in the California Mother Lode region between the 1850's and 1860's. Also included are hard to find details on the mineralization and locations of these mines, as well as how they were operated in those day. **8.5″ X 11″, 44 ppgs. Retail Price: $8.99**

<u>**Gold Rush on the Feather River**</u> - First published in 1924, this short publication by G.C. Mansfield sheds important light on the early history of gold mining on the Feather River. Included are rare insights into the first decade of gold mining and the early mining camps of the Feather River during the 1850's. 64 ppgs., 9.99

<u>**The Bodie Mining District of California**</u> - First published in 1986, it has been unavailable since those days and sheds important light on this famous mining area. Included are the history, characteristics and locations of numerous old mines around the ghost town of Bodie.
64 ppgs, 8.99

<u>**Geology and Mineral Resources of the Gasquet Quadrangle of California-Oregon**</u> - First published in 1953, it has been unavailable for over a century and sheds important light on the geological features and mineral resources of this portion of Northern California and Southern Oregon.
80 ppgs, 9.99

Alaska Mining Books

<u>**Ore Deposits of the Willow Creek Mining District, Alaska**</u> - Unavailable since 1954, this hard to find publication includes valuable insights into the Willow Creek Mining District near Hatcher Pass in Alaska. The publication includes insights into the history, geology and locations of the well known mines in the area, including the Gold Cord, Independence, Fern, Mabel, Lonesome, Snowbird, Schroff-O'Neil, High Grade, Marion Twin, Thorpe, Webfoot, Kelly-Willow, Lane, Holland and others. **8.5″ X 11″, 96 ppgs. Retail Price: $9.99**

<u>**The Juneau Gold Belt of Alaska**</u> - Unavailable since 1906, this hard to find publication includes valuable insights into the gold mines around Juneau, Alaska. The publication includes important details into the history, geology and locations of the well known gold mines and prospects in the area, including those around Windham Bay, Holkham Bay, Port Snettisham, on Grindstone and Rhine Creeks, Gold Creek, Douglas Island, Salmon Creek, Lemon Creek, Nugget Creek, from the Mendenhall River to Berners Bay, McGinnis Creek, Montana Creek, Peterson Creek, Windfall Creek, the Eagle River, Yankee Basin, Yankee Curve, Kowee Creek and elsewhere. Not only are gold placer mines included, but also hardrock gold mines. **8.5″ X 11″, 224 ppgs. Retail Price: $19.99**

<u>**Mining in the Jumbo Basin of Alaska**</u> - Unavailable since 1953, this hard to find publication includes valuable insights into the mines and geology of the Jumbo Basin. The publication includes important details into the history, geology and locations of the well known gold mines and prospects in the famous Jumbo Basin Mining Region of Alaska.
72 ppgs, 9.99

<u>**The Rampart Placer Gold Region of Alaska**</u> - Unavailable since 1906, this hard to find publication includes valuable insights into the placer gold mines of the Rampart Mining Region. The publication includes important details into the history, geology and locations of the well known gold mines and prospects in the famous Rampart Mining Region of Alaska.
78 ppgs, 10.99

Arizona Mining Books

<u>**Mines and Mining in Northern Yuma County Arizona**</u> - Originally published in 1911, this important publication on Arizona Mining has not been available for over a hundred years. Included are rare insights into the gold, silver, copper and quicksilver mines of Yuma County, Arizona together with hard to find maps and photographs. Some of the mines and mining districts featured include the Planet Copper Mine, Mineral Hill, the Clara Consolidated Mine, Viati Mine, Copper Basin prospect, Bowman Mine, Quartz King, Billy Mack, Carnation, the Wardwell and Osbourne, Valensuella Copper, the Mariquita, Colonial Mine, the French American, the New York-Plomosa, Guadalupe, Lead Camp, Mudersbach Copper Camp, Yellow Bird, the Arizona Northern (Salome Strike), Bonanza (Harqua Hala), Golden Eagle, Hercules, Socorro and others. **8.5″ X 11″, 144 ppgs. Retail Price: $11.99**

<u>**The Aravaipa and Stanley Mining Districts of Graham County Arizona**</u> - Originally published in 1925, this important publication on Arizona Mining has not been available for nearly ninety years. Included are rare insights into the gold and silver mines of these two important mining districts, together with hard to find maps. **8.5″ X 11″, 140 ppgs. Retail Price: $11.99**

Gold in the Gold Basin and Lost Basin Mining Districts of Mohave County, Arizona - This volume contains rare insights into the geology and gold mineralization of the Gold Basin and Lost Basin Mining Districts of Mohave County, Arizona that will be of benefit to miners and prospectors. Also included is a significant body of information on the gold mines and prospects of this portion of Arizona. This volume is lavishly illustrated with rare photos and mining maps. **8.5" X 11", 188 ppgs. Retail Price: $19.99**

Mines of the Jerome and Bradshaw Mountains of Arizona - This important publication on Arizona Mining has not been available for ninety years. This volume contains rare insights into the geology and ore deposits of the Jerome and Bradshaw Mountains of Arizona that will be of benefit to miners and prospectors who work those areas. Included is a significant body of information on the mines and prospects of the Verde, Black Hills, Cherry Creek, Prescott, Walker, Groom Creek, Hassayampa, Bigbug, Turkey Creek, Agua Fria, Black Canyon, Peck, Tiger, Pine Grove, Bradshaw, Tintop, Humbug and Castle Creek Mining Districts. This volume is lavishly illustrated with rare photos and mining maps. **8.5" X 11", 218 ppgs. Retail Price: $19.99**

The Ajo Mining District of Pima County Arizona - This important publication on Arizona Mining has not been available for nearly seventy years. This volume contains rare insights into the geology and mineralization of the Ajo Mining District in Pima County, Arizona and in particular the famous New Cornelia Mine. **8.5" X 11", 126 ppgs. Retail Price: $11.99**

Mining in the Santa Rita and Patagonia Mountains of Arizona - Originally published in 1915, this important publication on Arizona Mining has not been available for nearly a century. Included are rare insights into hundreds of gold, silver, copper and other mines in this famous Arizona mining area. Details include the locations, geology, history, production and other facts of the mines of this region. **8.5" X 11", 394 ppgs. Retail Price: $24.99**

Mining in the Bisbee Quadrangle of Arizona - Originally published in 1906, this important publication on Arizona Mining has not been available for nearly a century. Included are rare insights into hundreds of gold, silver, copper and other mines in this famous Arizona mining area. Details include the locations, geology, history, production and other facts of the mines of this important mining region. **8.5" X 11", 188 ppgs. Retail Price: $14.99**

Placer Gold Mining in Arizona - Unavailable since 1922, this hard to find publication includes valuable insights into the placer gold mines of the Arizona. Originally released as "Placer Gold of Arizona", despite its small size, this publication includes important details into the history, geology and locations of the well known placer gold mines and prospects in the State of Arizona. **48 ppgs, 8.99**

Gold and Copper Mining near Payson, Arizona - Written in 1915, this hard to find publication includes valuable insights into the gold and copper mining industry of Arizona. Highlighted here are the gold and copper mines near Payson, Arizona. **68 ppgs, 8.99**

Lode Gold Mining in Arizona - Unavailable since 1934, this hard to find publication, originally released as "Arizona Lode Gold Mines and Gold Mining" includes valuable insights into the gold mining industry of Arizona. Included are valuable insights into over 150 hardrock gold mines in over 30 different mining districts in Arizona. **278 ppgs, 21.99**

Mining in the Dragoon Quadrangle of Cochise County, Arizona - Unavailable since 1964, this hard to find publication includes valuable insights into the mines of the Dragoon Quadrangle Mining Region. The publication includes important details into the history, geology and locations of the well known mines and prospects in this famous mining region of Arizona. **224 ppgs., 19.99**

Directory of Operating Mines in Arizona in 1915 - Unavailable since 1916, this hard to find publication includes valuable insights into the mines of Arizona. This small publication includes a complete list of the mines that were operating in the State of Arizona during 1915 and includes details such as general location, owners and some basic facts about each mining operation. **52 ppgs. 8.99**

Arizona Ore Deposits - Unavailable since 1938, this hard to find publication includes valuable insights into some ore deposits of Arizona. Included are valuable insights into the formation and characteristics of valuable ore deposits in the Jerome, Miami, Inspiration, Clifton, Morenci, Ray, Ajo, Eureka, Tombstone and Magma mining districts. Included are details into some of the major gold, silver and copper mines of these important Arizona mining areas. **160 ppgs, 14.99**

Montana Mining Books

A History of Butte Montana: The World's Greatest Mining Camp - First published in 1900 by H.C. Freeman, this important publication sheds a bright light on one of the most important mining areas in the history of The West. Together with his insights, as well as rare photographs of the periods, Harry Freeman describes Butte and its vicinity from its early beginnings, right up to its flush years when copper flowed from its mines like a river. At the time of publication, Butte, Montana was known worldwide as "The Richest Mining Spot On Earth" and produced not only vast amounts of copper, but also silver, gold and other metals from its mines. Freeman illustrates, with great detail, the most important mines in the vicinity of Butte, providing rare details on their owners, their history and most importantly, how the mines operated and how their treasures were extracted. Of particular interest are the dozens of rare photographs that depict mines such as the famous Anaconda, the Silver Bow, the Smoke House, Moose, Paulin, Buffalo, Little Minah, the Mountain Consolidated, West Greyrock, Cora, the Green Mountain, Diamond, Bell, Parnell, the Neversweat, Nipper, Original and many others.   8.5" X 11", 142 ppgs. Retail Price: $12.99

The Butte Mining District of Montana - This important publication on Montana Mining has not been available for over a century. Included are rare insights into the gold, copper and silver mines of Butte, Montana together with hard to find maps and photographs. Some of the topics include the early history of gold, silver and copper mining in the Butte area, insight into the geology of its mining areas, the local distribution of gold, silver and copper ores, as well their composition and how to identify them. Also included are detailed facts about the mines in the Butte Mining District, including the famous Anaconda Mine, Gagnon, Parrot, Blue Vein, Moscow, Poulin, Stella, Buffalo, Green Mountain, Wake Up Jim, the Diamond-Bell Group, Mountain Consolidated, East Greyrock, West Greyrock, Snowball, Corra, Speculator, Adirondack, Miners Union, the Jessie-Edith May Group, Otisco, Iduna, Colorado, Lizzie, Cambers, Anderson, Hesperus, Preferencia and dozens of others. **8.5" X 11", 298 ppgs. Retail Price: $24.99**

Mines of the Helena Mining Region of Montana - This important publication on Montana Mining has not been available for over a century. Included are rare insights into the gold, copper and silver mines of the vicinity of Helena, Montana, including the Marysville Mining District, Elliston Mining District, Rimini Mining District, Helena Mining District, Clancy Mining District, Wickes Mining District, Boulder and Basin Mining Districts and the Elkhorn Mining District. Some of the topics include the early history of gold, silver and copper mining in the Helena area, insight into the geology of its mining areas, the local distribution of gold, silver and copper ores, as well their composition and how to identify them. Also included are detailed facts, history, geology and locations of over one hundred gold, silver and copper mines in the area . **8.5" X 11", 162 ppgs, Retail Price: $14.99**

Mines and Geology of the Garnet Range of Montana - This important publication on Montana Mining has not been available for over a century. Included are rare insights into the gold, copper and silver mines of the vicinity of this important mining area of Montana. Some of the topics include the early history of gold, silver and copper mining in the Garnet Mountains, insight into the geology of its mining areas, the local distribution of gold, silver and copper ores, as well their composition and how to identify them. Also included are detailed facts, history, geology and locations of numerous gold, silver and copper mines in the area . **8.5" X 11", 100 ppgs, Retail Price: $11.99**

Mines and Geology of the Philipsburg Quadrangle of Montana - This important publication on Montana Mining has not been available for over a century. Included are rare insights into the gold, copper and silver mines of the vicinity of this important mining area of Montana. Some of the topics include the early history of gold, silver and copper mining in the Philipsburg Quadrangle, insight into the geology of its mining areas, the local distribution of gold, silver and copper ores, as well their composition and how to identify them. Also included are detailed facts, history, geology and locations of over one hundred gold, silver and copper mines in the area **8.5" X 11", 290 ppgs, Retail Price: $24.99**

Geology of the Marysville Mining District of Montana - Included are rare insights into the mining geology of the Marysville Mining District. Some of the topics include the early history of gold, silver and copper mining in the area, insight into the geology of its mining areas, the local distribution of gold, silver and copper ores, as well their composition and how to identify them. Also included are detailed facts, history, geology and locations of gold, silver and copper mines in the area **8.5" X 11", 198 ppgs, Retail Price: $19.99**

The Geology and Mines of Northern Idaho and North Western Montana- See listing under Idaho.

The History of Gold Dredging in Montana - Unavailable since 1916, this important publication was originally published by the Us Bureau of Mines and has been unavailable for a century. A century and more ago, giant dredging machines dug in Montana's rivers and creeks in search of illusive golden riches. First appearing in California in the 1850's, gold dredges finally reached their peak of development in Siberia and New Zealand before becoming popular again in the United States. This book offers a unique historical perspective on the gold dredges that once operated in Montana. This book on Montana mining history is lavishly illustrated with dozens of rare historic photos gold dredges that once operated in Montana, as well as hard to locate plans on how these dredges were designed. 120 ppgs., 11.99

Nevada Mining Books

The Bull Frog Mining District of Nevada - Unavailable since 1910, this publication was originally compiled by the United States Department of Interior. This volume also includes important insights into the geologic formations, faults and other aspects of economic geology in this Nevada mining district. Of particular interest are the fine details on many mines in the area, including their locations, histories, development and mineralization. Some of the mines featured include the National Bank Mine, Providence, Gibraltor, Tramps, Denver, Original Bullfrog, Gold Bar, Mayflower, Homestake-King and other mines and prospects. **8.5" X 11", 152 ppgs, Retail Price: $14.99**

History of the Comstock Lode - Unavailable since 1876, this publication was originally released by John Wiley & Sons. This volume also includes important insights into the famous Comstock Lode of Nevada that represented the first major silver discovery in the United States. During its spectacular run, the Comstock produced over 192 million ounces of silver and 8.2 million ounces of gold. Not only did the Comstock result in one of the largest mining rushes in history and yield immense fortunes for its owners, but it made important contributions to the development of the State of Nevada, as well as neighboring California. Included here are important details on not only the early development and history of the Comstock, but also rare early insight into its mines, ore and its geology.8.5" X 11", 244 ppgs, Retail Price: $19.99

The Pioche Mining District of Nevada - First published in 1932, it has been unavailable for over a century and sheds important light on the mining history of Nevada. Some of the topics include the history of mining in this district, as well as the characteristics of its mineral and ore deposits. Also included are insights into the history, production, characteristics and locations of numerous mines in the area. Some of the mines include the Combined Metals, Pioche, Ely Valley, No. 10, Poorman, Wide Awake, Alps, Prince, Virginia Louise, Half Moon, Abe Lincoln, Fairview, Bristol Silver, National, Vesuvius, Inman, Tempest, Hillside, Jackrabbit, Lucky Star, Fortuna, Mendha, Manhattan, Hamburg, Comet, Lyndon and others. 108 ppgs 10.99

The Yerington Mining District of Nevada - First published in 1932, it has been unavailable for over a century and sheds important light on the mining history of Nevada. Some of the topics include the history of mining in this district, as well as the characteristics of its mineral and ore deposits. Also included are insights into the history, production, characteristics and locations of numerous mines in the area. Some of the mines include the Bluestone, Mason Valley, Malachite, McConnell, Greenwood, Western Nevada, Ludwig, Douglas Hill, Casting Copper, Montana-Yerington, Empire, Jim Beatty, Terry and McFarland, Blue Jay and others. 92 ppgs, 10.99

The Genesis of the Ores of Tonopah Nevada - Unavailable since 1918, this hard to find publication includes valuable insights into the gold mines around Tonopah, Nevada. The publication includes important details into the geology of mines in the Tonopah Mining District of Nevada. 90 ppgs, 10.99

Mining Camps of Elko, Lander and Eureka Counties Nevada - Unavailable since 1910, this hard to find publication includes valuable insights into the mining camps of Elko, Lander and Eureka Counties, Nevada. The publication includes important details into the history of mines and mining in these three Nevada counties. 154 ppgs, 12.99

Ore Deposits of the Bullfrog Quadrangle - Unavailable since 1964 and released as "Geology of Bullfrog Quadrangle and Ore Deposits Related to Bullfrog Hills Caldera, Nye County, Nevada and Inyo County, California". The publication includes important details into the geology of mines in the Bullfrog Quadrangle of Nye County, Nevada and Inyo County, California. 52 ppgs, 9.99

Mining in Eureka County Nevada - Unavailable since 1879, this hard to find publication includes valuable insights into the early mining history off Eureka County, Nevada. The publication includes important details into the early history of the mines of Eureka County, as well as their development, production and how their ores were treated. Also included are details on the 1872 Mining Act, as well as the local rules, regulations and customs of the miners in Eureka County.134 ppgs, 12.99

Colorado Mining Books

Ores of The Leadville Mining District - Unavailable since 1926, this publication was originally compiled by the United States Department of Interior. This volume also includes important insights into the ores and mineralization of the Leadville Mining District in Colorado. Topics include historic ore prospecting methods, local geology, insights into ore veins and stockworks, the local trend and distribution of ore channels, reverse faults, shattered rock above replacement ore bodies, mineral enrichment in oxidized and sulphide zones and more. **8.5″ X 11″, 66 ppgs, Retail Price: $8.99**

Mining in Colorado - Unavailable since 1926, this publication was originally compiled by the United States Department of Interior. This volume also includes important insights into the mining history of Colorado from its early beginnings in the 1850's right up to the mid 1920's. Not only is Colorado's gold mining heritage included, but also its silver, copper, lead and zinc mining industry. Each mining area is treated separately, detailing the development of Colorado's mines on a county by county basis. **8.5″ X 11″, 284 ppgs, Retail Price: $19.99**

Gold Mining in Gilpin County Colorado - Unavailable since 1876, this publication was originally compiled by the Register Steam Printing House of Central City, Colorado. A rare glimpse at the gold mining history and early mines of Gilpin County, Colorado from their first discovery in the 1850's up to the "flush years" of the mid 1870's. Of particular interest is the history of the discovery of gold in Gilpin County and details about the men who made those first strikes. Special focus is given to the early gold mines and first mining districts of the area, many of which are not detailed in other books on Colorado's gold mining history. **8.5″ X 11″, 156 ppgs, Retail Price: $12.99**

Mining in the Gold Brick Mining District of Colorado - Important insights into the history of the Gold Brick Mining District, as well as its local geography and economic geology. Also included are the histories and locations of historic mines in this important Colorado Mining District, including the Cortland, Carter, Raymond, Gold Links, Sacramento, Bassick, Sandy Hook, Chronicle, Grand Prize, Chloride, Granite Mountain, Lucille, Gray Mountain, Hilltop, Maggie Mitchell, Silver Islet, Revenue, Roosevelt, Carbonate King and others. In addition to hardrock mining, are also included are details on gold placer mining in this portion of Colorado. **8.5″ X 11″, 140 ppgs, Retail Price: $12.99**

Ore Deposits of the London Fault of Colorado - First published in 1941, it has been unavailable since those days and sheds important light on the mines and mineral deposits of the London Fault in Central Colorado's Alma Mining District. This publication sheds important light on the gold veins and lead-silver deposits of the Alma Mining District. Included are geologic details on the London Mine, American Mine, Havigorst Tunnel, Ophir Mine, Mosher Tunnel, London-Butte Mine, Venture Shaft, Hard-To-Beat Mine, Oliver Twist Tunnel, Sacramento Mine, Mudsill Mine, Sherwood Mine, Wagner, Barcoe Tunnel and other mines in this important mining region. 110 ppgs., 10.99

The Mines of Colorado - First published in 1867, it has been unavailable since those days and sheds important light on Colorado's early mining history. Written shortly after the events took place, this publication sheds important light on the Pike's Peak Gold Rush, the discovery of gold on Ralston Creek and Dry Creek in the 1850's, as well as details on the first wave of miners into Colorado and their trials and tribulations as they crossed the Great Plains. Also included are details on early discoveries of lode gold in the mountainous regions of Colorado, details on the early mines hardrock and placer mines, and much more. It is a veritable treasure trove on Colorado's early mining history and will be of great importance to anyone who is interested in the mining of gold or other minerals in Colorado, as well as those interested in the history of the state. 478 ppgs., 29.99

The La Plata Mining District of Colorado - Originally titled "Geology and Ore Deposits in the Vicinity of the La Plata District of Colorado" and first published in 1949, it has been unavailable since those days and sheds important light on the mines and mineral deposits of the La Plata Mining District of Colorado. 214 ppgs., 19.99

Washington Mining Books

The Republic Mining District of Washington - Unavailable since 1910, this important publication was originally published by the Washington Geologic Survey and has been unavailable for a century. Topics include the geology, rock formations and the formation of ore deposits in this important mining area of Washington State. Also included are hard to find details on the geology, history and locations of dozens of mines in the area. Some of the mines featured include the New Republic Mine, Ben Hur, Morning Glory, the South Republic Mine, Quilp, Surprise, Black Tail, Lone Pine, San Poil, Mountain Lion, Tom Thumb, Elcaliph and many others. **8.5" X 11", 94 ppgs, Retail Price: $10.99**

The Myers Creek and Nighthawk Mining Districts of Washington - Unavailable since 1911, this important publication was originally published by the Washington Geologic Survey and has been unavailable for a century. Topics include the geology, rock formations and the formation of ore deposits in these important mining areas of Washington State. Also included are hard to find details on the geology, history and locations of dozens of mines in the area. Some of the mines featured include the Grant Mine, Monterey, Nip and Tuck, Myers Creek, Number Nine, Neutral, Rainbow, Aztec, Crystal Butte, Apex, Butcher Boy, Molson, Mad River, Olentangy, Delate, Kelsey, Golden Chariot, Okanogan, Ohio, Forty-Ninth Parallel, Nighthawk, Favorite, Little Chopaka, Summit, Number One, California, Peerless, Caaba, Prize Group, Ruby, Mountain Sheep, Golden Zone, Rich Bar, Similkameen, Kimberly, Triune, Hiawatha, Trinity, Hornsilver, Maquae, Bellevue, Bullfrog, Palmer Lake, Ivanhoe, Copper World and many others. **8.5" X 11", 136 ppgs, Retail Price: $12.99**

The Blewett Mining District of Washington - Unavailable since 1911, this important publication was originally published by the Washington Geologic Survey and has been unavailable for a century. Topics include the geology, rock formations and the formation of ore deposits in this important mining area of Washington State. Also included are hard to find details on the geology, history and locations of dozens of mines in the area. Some of the mines featured include the Washington Meteor, Alta Vista, Pole Pick, Blinn, North Star, Golden Eagle, Tip Top, Wilder, Golden Guinea, Lucky Queen, Blue Bell, Prospect, Homestake, Lone Rock, Johnson, and others. **8.5" X 11", 134 ppgs, Retail Price: $12.99**

Silver Mining In Washington - Unavailable since 1955, this important publication was originally published by the Washington Geologic Survey. Featured are the hard to find locations and details pertaining to Washington's silver mines. **8.5" X 11", 180 ppgs, Retail Price: $15.99**

The Mines of Snohomish County Washington - Unavailable since 1942, this important publication was originally published by the Washington Geologic Survey and has been unavailable for seventy years. Featured are details on a large number of gold, silver, copper, lead and other metallic mineral mines. Included are the locations of each historic mine, along with information on the commodity produced. **8.5" X 11", 98 ppgs, Retail Price: $10.99**

The Mines of Chelan County Washington - Unavailable since 1943, this important publication was originally published by the Washington Geologic Survey and has been unavailable for seventy years. Featured are details on a large number of gold, silver, copper, lead and other metallic mineral mines. Included are the locations of each historic mine, along with information on the commodity. **8.5" X 11", 88 ppgs, Retail Price: $9.99**

Metal Mines of Washington - Unavailable since 1921, this important publication was originally published by the Washington Geologic Survey and has been unavailable for nearly ninety years. Widely considered a masterpiece on the Washington Mining Industry, "Metal Mines of Washington" sheds light on the important details of Washington's early mining years. Featured are details on hundreds of gold, silver, copper, lead and other metallic mineral mines. Included are hard to find details on the mineral resources of this state, as well as the locations of historic mines. Lavishly illustrated with maps and historic photos and complete with a glossary to explain any technical terms found in the text, this is one of the most important works on mining in the State of Washington. No prospector or miner should be without it if they are interested in mining in Washington. **8.5" X 11", 396 ppgs, Retail Price: $24.99**

Gem Stones In Washington - Unavailable since 1949, this important publication was originally published by the Washington Geologic Survey and has been unavailable since first published. Included are details on where to find naturally occurring gem stones in the State of Washington, including quartz crystal, amethyst, smoky quartz, milky quartz, agates, bloodstone, carnelian, chert, flint, jasper, onyx, petrified wood, opal, fire opal, hyalite and others. **8.5" X 11", 54 ppgs, Retail Price: $8.99**

The Covada Mining District of Washington - Unavailable since 1913, this important publication was originally published by the Washington Geologic Survey and has been unavailable for a century. Topics include the geology, rock formations and the formation of ore deposits in this important mining area of Washington State. Also included are hard to find details on the geology, history and locations of dozens of mines in the area. Some of the mines featured include the Admiral, Advance, Algonkian, Big Bug, Big Chief, Big Joker, Black Hawk, Black Tail, Black Thorn, Captain, Cherokee Strip, Colorado, Dan Patch, Dead Shot, Etta, Good Ore, Greasy Run, Great Scott, Idora, IXL, Jay Bird, Kentucky Bell, King Solomon, Laurel, Laura S, Little Jay, Meteor, Neglected, Northern Light, Old Nell, Plymouth Rock, Polaris, Quandary, Reserve, Shoo Fly, Silver Plume, Three Pines, Vernie, White Rose and dozens of others. **8.5" X 11", 114 ppgs, Retail Price: $10.99**

The Index Mining District of Washington - Unavailable since 1912, this important publication was originally published by the Washington Geologic Survey and has been unavailable for a century. Topics include the geology, rock formations and the formation of ore deposits in this important mining area of Washington State. Also included are hard to find details on the geology, history and locations of dozens of mines in the area. Some of the mines featured include the Sunset, Non-Pareil, Ethel Consolidated, Kittaning, Merchant, Homestead, Co-operative, Lost Creek, Uncle Sam, Calumet, Florence-Rae, Bitter Creek, Index Peacock, Gunn Peak, Helena, North Star, Buckeye. Copper Bell, Red Cross and others. **8.5″ X 11″, 114 ppgs, Retail Price: $11.99**

Mining & Mineral Resources of Stevens County Washington - Unavailable since 1920, this important publication was originally published by the Washington Geologic Survey and has been unavailable for a century. Topics include the geology, rock formations and the formation of ore deposits in these important mining areas of Washington State. Also included are hard to find details on the geology, history and locations of hundreds of mines in the area. **8.5″ X 11″, 372 ppgs, Retail Price: $24.99**

The Mines and Geology of the Loomis Quadrangle Okanogan County, Washington - Unavailable since 1972, this important publication was originally published by the Washington Geologic Survey and has been unavailable for a century. Topics include the geology, rock formations and the formation of ore deposits in this important mining area of Washington State. Also included are hard to find details on the geology, history and locations of dozens of gold, copper, silver and other mines in the area. **8.5″ X 11″, 150 ppgs, Retail Price: $12.99**

The Conconully Mining District of Okanogan County Washington - Unavailable since 1973, this important publication was originally published by the Washington Geologic Survey and has been unavailable for a century. Topics include the geology, rock formations and the formation of ore deposits in this important mining area of Washington State, which also includes Salmon Creek, Blue Lake and Galena. Also included are hard to find details on the geology, mining history and locations of dozens of mines in the area. Some of the mines include Arlington, Fourth of July, Sonny Boy, First Thought, Last Chance, War Eagle-Peacock, Wheeler, Mohawk, Lone Star, Woo Loo Moo Loo, Keystone, Hughes, Plant-Callahan, Johnny Boy, Leuena, Gubser, John Arthur, Tough Nut, Homestake, Key and many others **8.5″ X 11″, 68 ppgs, Retail Price: $8.99**

Wyoming Mining Books

Mining in the Laramie Basin of Wyoming - Unavailable since 1909, this publication was originally compiled by the United States Department of Interior. Also included are insights into the mineralization and other characteristics of this important mining region, especially in regards to coal, limestone, gypsum, bentonite clay, cement, sand, clay and copper. **8.5″ X 11″, 104 ppgs, Retail Price: $11.99**

New Mexico Mining Books

The Mogollon Mining District of New Mexico - Unavailable since 1927, this important publication was originally published by the US Department of Interior and has been unavailable for 80 years. Topics include the geology, rock formations and the formation of ore deposits in this important mining area in New Mexico. Of particular focus is information on the history and production of the ore deposits in this area, their form and structure, vein filling, their paragenesis, origins and ore shoots, as well as oxidation and supergene enrichment. Also included are hard to find details, including the descriptions and locations of numerous gold, silver and other types of mines, including the Eureka, Pacific, South Alpine, Great Western, Enterprise, Buffalo, Mountain View, Floride, Gold Dust, Last Chance, Deadwood, Confidence, Maud S., Deep Down, Little Fanney, Trilby, Johnson, Alberta, Comet, Golden Eagle, Cooney, Queen, the Iron Crown, Eberle, Clifton, Andrew Jackson mine, Mascot and others. **8.5″ X 11″, 144 ppgs, Retail Price: $12.99**

The Percha Mining District of Kingston New Mexico - Unavailable since 1883, this important publication was originally published by the Kingston Tribune and has been unavailable for over one hundred and thirty five years. Having been written during the earliest years of gold and silver mining in the Percha Mining District, unlike other books on the subject, this work offers the unique perspective of having actually been written while the early mining history of this area was still being made. In fact, the work was written so early in the development of this area that many of the notable mines in the Percha District were less than a few years old and were still being operated by their original discoverers with the same enthusiasm as when they were first located. Included are hard to find details on the very earliest gold and silver mines of this important mining district near Kingston in Sierra County, New Mexico. **8.5″ X 11″, 68 ppgs, Retail Price: $9.99**

East Coast Mining Books

The Gold Fields of the Southern Appalachians - Unavailable since 1895, this important publication was originally published by the US Department of Interior and has been unavailable for nearly 120 years. Topics include the geology, rock formations and the formation of ore deposits in this important mining area of the American South. Of particular focus is information on the history and statistics of the ore deposits in this area, their form and structure and veins. Also included are details on the placer gold deposits of the region. The gold fields of the Georgian Belt, Carolinian Belt and the South Mountain Mining District of North Carolina are all treated in descriptive detail. Included are hard to find details, including the descriptions and locations of numerous gold mines in Georgia, North Carolina and elsewhere in the American South. Also included are details on the gold belts of the British Maritime Provinces and the Green Mountains. **8.5" X 11", 104 ppgs, Retail Price: $9.99**

Gold Rush Tales Series

Millions in Siskiyou County Gold - In this first volume of the "Gold Rush Tales" series, leading mining historian and editor Kerby Jackson, introduces us to the story of how millions of dollars worth of gold was discovered in Siskiyou County during the California Gold Rush. Lavishly illustrated with photos from the 19th Century, this hard to find information was first published in 1897 and sheds important light onto the gold rush era in Siskiyou County, California and the experiences of the men who dug for the gold and actually found it. **8.5" X 11", 82 ppgs, Retail Price: $9.99**

The California Rand in the Days of '49 - In this second volume of the "Gold Rush Tales" series, leading mining historian and editor Kerby Jackson, introduces us to four tales from the California Gold Rush. Lavishly illustrated with photos from the 19th Century, this hard to find information was first published in 1890's and includes the stories of "California's Rand", details about Chinese miners, how one early miner named Baker struck it rich and also the story of Alphonzo Bowers, who invented the first hydraulic gold dredge. **8.5" X 11", 54 ppgs, Retail Price: $9.99**

More Mining Books

Prospecting and Developing A Small Mine - Topics covered include the classification of varying ores, how to take a proper ore sample, the proper reduction of ore samples, alluvial sampling, how to understand geology as it is applied to prospecting and mining, prospecting procedures, methods of ore treatment, the application of drilling and blasting in a small mine and other topics that the small scale miner will find of benefit. **8.5" X 11", 112 ppgs, Retail Price: $11.99**

Timbering For Small Underground Mines - Topics covered include the selection of caps and posts, the treatment of mine timbers, how to install mine timbers, repairing damaged timbers, use of drift supports, headboards, squeeze sets, ore chute construction, mine cribbing, square set timbering methods, the use of steel and concrete sets and other topics that the small underground miner will find of benefit. This volume also includes twenty eight illustrations depicting the proper construction of mine timbering and support systems that greatly enhance the practical usability of the information contained in this small book. **8.5" X 11", 88 ppgs. Retail Price: $10.99**

Timbering and Mining - A classic mining publication on Hard Rock Mining by W.H. Storms. Unavailable since 1909, this rare publication provides an in depth look at American methods of underground mine timbering and mining methods. Topics include the selection and preservation of mine timbers, drifting and drift sets, driving in running ground, structural steel in mine workings, timbering drifts in gravel mines, timbering methods for driving shafts, positioning drill holes in shafts, timbering stations at shafts, drainage, mining large ore bodies by means of open cuts or by the "Glory Hole" system, stoping out ore in flat or low lying veins, use of the "Caving System", stoping in swelling ground, how to stope out large ore bodies, Square Set timbering on the Comstock and its modifications by California miners, the construction of ore chutes, stoping ore bodies by use of the "Block System", how to work dangerous ground, information on the "Delprat System" of stoping without mine timbers, construction and use of headframes and much more. This volume provides a reference into not only practical methods of mining and timbering that may be employed in narrow vein mining by small miners today, but also rare insights into how mines were being worked at the turn of the 19th Century. **8.5" X 11", 288 ppgs. Retail Price: $24.99**

A Study of Ore Deposits For The Practical Miner - Mining historian Kerby Jackson introduces us to a classic mining publication on ore deposits by J.P. Wallace. First published in 1908, it has been unavailable for over a century. Included are important insights into the properties of minerals and their identification, on the occurrence and origin of gold, on gold alloys, insights into gold bearing sulfides such as pyrites and arsenopyrites, on gold bearing vanadium, gold and silver tellurides, lead and mercury tellurides, on silver ores, platinum and iridium, mercury ores, copper ores, lead ores, zinc ores, iron ores, chromium ores, manganese ores, nickel ores, tin ores, tungsten ores and others. Also included are facts regarding rock forming minerals, their composition and occurrences, on igneous, sedimentary, metamorphic and intrusive rocks, as well as how they are geologically disturbed by dikes, flows and faults, as well as the effects of these geologic actions and why they are important to the miner. Written specifically with the common miner and prospector in mind, the book will help to unlock the earth's hidden wealth for you and is written in a simple and concise language that anyone can understand. **8.5″ X 11″, 366 ppgs. Retail Price: $24.99**

Mine Drainage - Unavailable since 1896, this rare publication provides an in depth look at American methods of underground mine drainage and mining pump systems. This volume provides a reference into not only practical methods of mining drainage that may be employed in narrow vein mining by small miners today, but also rare insights into how mines were being worked at the turn of the 19th Century. **8.5″ X 11″, 218 ppgs. Retail Price: $24.99**

Fire Assaying Gold, Silver and Lead Ores - Unavailable since 1907, this important publication was originally published by the Mining and Scientific Press and was designed to introduce miners and prospectors of gold, silver and lead to the art of fire assaying. Topics include the fire assaying of ores and products containing gold, silver and lead; the sampling and preparation of ore for an assay; care of the assay office, assay furnaces; crucibles and scorifiers; assay balances; metallic ores; scorification assays; cupelling; parting' crucible assays, the roasting of ores and more. This classic provides a time honored method of assaying put forward in a clear, concise and easy to understand language that will make it a benefit to even beginners. **8.5″ X 11″, 96 ppgs. Retail Price: $11.99**

Methods of Mine Timbering - Originally published in 1896, this important publication on mining engineering has not been available for nearly a century. Included are rare insights into historical methods of timbering structural support that were used in underground metal mines during the California that still have a practical application for the small scale hardrock miner of today. **8.5″ X 11″, 94 ppgs. Retail Price: $10.99**

The Enrichment of Copper Sulfide Ores - First published in 1913, it has been unavailable for over a century. Topics include the definition and types of ore enrichment, the oxidation of copper ores, the precipitation of metallic sulfides. Also included are the results of dozens of lab experiments pertaining to the enrichment of sulfide ores that will be of interest to the practical hard rock mine operator in his efforts to release the metallic bounty from his mine's ore. **8.5″ X 11″, 92 ppgs. Retail Price: $9.99**

A Study of Magmatic Sulfide Ores - Unavailable since 1914, this rare publication provides an in depth look at magmatic sulfide ores. Some of the topics included are the definition and classification of magmatic ores, descriptions of some magmatic sulfide ore deposits known at the time of publication including copper and nickel bearing pyrrohitic ore bodies, chalcopyrite-bornite deposits, pyritic deposits, magnetite-ileminite deposits, chromite deposits and magmatic iron ore deposits. Also included are details on how to recognize these types of ore deposits while prospecting for valuable hardrock minerals. **8.5″ X 11″, 138 ppgs. Retail Price: $11.99**

The Cyanide Process of Gold Recovery - Unavailable since 1894 and released under the name "The Cyanide Process: Its Practical Application and Economical Results", this rare publication provides an in depth look at the early use of cyanide leaching for gold recovery from hardrock mine ores. This volume provides a reference into the early development and use of cyanide leaching to recover gold. **8.5″ X 11″, 162 ppgs. Retail Price: $14.99**

California Gold Milling Practices - Unavailable since 1895 and released under the name "California Gold Practices", this rare publication provides an in depth look at early methods of milling used to reduce gold ores in California during the late 19th century. This volume provides a reference into the early development and use of milling equipment during the earliest years of the California Gold Rush up to the age of the Industrial Revolution. Much of the information still applies today and will be of use to small scale miners engaging in hardrock mining. **8.5″ X 11″, 104 ppgs. Retail Price: $10.99**

Leaching Gold and Silver Ores With The Plattner and Kiss Processes - Mining historian Kerby Jackson introduces us to a classic mining publication on the evaluation and examination of mines and prospects by C.H. Aaron. First published in 1881, it has been unavailable for over a century and sheds important light on the leaching of gold and silver ores with the Plattner and Kiss processes. **8.5″ X 11″, 204 ppgs. Retail Price: $15.99**

The Metallurgy of Lead and the Desilverization of Base Bullion - First published in 1896, it has been unavailable for over a century and sheds important light on the the recovery of silver from lead based ores. Some of the topics include the properties of lead and some of its compounds, lead ores such as galenite, anglesite, cerussite and others, the distribution of lead ores throughout the United States and the sampling and assaying of lead ores. Also covered is the metallurgical treatment of lead ores, as well as the desilverization of lead by the Pattinson Process and the Parkes Process. Hofman's text has long been considered one of the most important early works on the recovery of silver from lead based ores. 8.5" X 11", 452 ppgs. **Retail Price: $29.99**

Ore Sampling For Small Scale Miners - First published in 1916, it has been unavailable for over a century and sheds important light on historic methods of ore sampling in hardrock mines. Topics include how to take correct ore samples and the conditions that affect sampling, such as their subdivision and uniformity. Particular detail is given to methods of hand sampling ore bodies by grab sample, pipe sample and coning, as well as sampling by mechanical methods. Also given are insights into the screening, drying and grinding processes to achieve the most consistent sample results and much more. 8.5" X 11", 124 ppgs. **Retail Price: $12.99**

The Extraction of Silver, Copper and Tin from Ores - First published in 1896, it has been unavailable for over a century and sheds important light on how historic miners recovered silver, copper and tin from their mining operations. The book is split into three sections, including a discussion on the Lixiviation of Silver Ores, the mining and treatment of copper ores as practiced at Tharsis, Spain and the smelting of tin as it was practiced by metallurgists at Pulo Brani, Singapore. Also included is an overview and analysis of these historic metal recovery methods that will be of benefit to those interested in the extraction of silver, copper and tin from small mines. 8.5" X 11", 118 ppgs. **Retail Price: $14.99**

The Roasting of Gold and Silver Ores - First published in 1880, it has been unavailable for over a century and sheds important light on how historic miners recovered gold and silver rom their mining operations. Topics include details on the most important silver and free milling gold ores, methods of desulphurization of ores, methods of deoxidation, the chlorination of ores, methods and details on roasting gold and silver ores, notes on furnaces and more. Also included are details on numerous methods of gold and silver recovery, including the Ottokar Hofman's Process, the Patera Process, Kiss Process, Augustin Process, Ziervogel Process and others. 8.5" X 11", 178 ppgs. **Retail Price: $19.99**

The Examination of Mines and Prospects - First published in 1912, it has been unavailable for over a century and sheds important light on how to examine and evaluate hardrock mines, prospects and lode mining claims. Sections include Mining Examinations, Structural Geology, Structural Features of Ore Deposits, Primary Ores and their Distribution, Types of Primary Ore Deposits, Primary Ore Shoots, The Primary Alteration of Wall Rocks, Alterations by Surface Agencies, Residual Ores and their Distribution, Secondary Ores and Ore Shoots and Vein Outcrops. This hard to find information is a must for those who are interested in owning a mine or who already own a lode mining claim and wish to succeed at quartz mining. 8.5" X 11", 250 ppgs. **Retail Price: $19.99**

Garnets: Their Mining, Milling and Utilization - First published in 1925, it has been unavailable since those days and sheds important light on the mining, milling and utilization of garnets. Included are details on the characteristics of garnets, where they are found and how they were mined. 78 ppgs, 10.99

Gemstones and Precious Stones of North America - Leading mining historian Kerby Jackson introduces us to a classic mining publication on the gems and precious stones of the United States, Canada and mexico. First published in 1890, it has been unavailable since those days and sheds important light on the gems and precious stones that may be found in North America. Included are chapters on diamonds, corundum, sapphire, ruby, topaz, emerald, disapore, spinel, turquoise, tourmaline, garnets, beyrl, peridot, zircon, quartz crystals, feldspars, pearls and many others. Included are details on where these gems and precious stones may be found throughout North America, as well as their characteristics. 360 ppgs, 24.99

Mining Camps and Mining Districts - First released in 1885 by Charles Howard Shinn under the title "Mining Camps: A Study in American Frontier Government", this publication offers a unique look at how early gold miners established their own forms of representative government during the California Gold Rush. Drawing on the the early mining codes of medieval German miners in the Harz Mountains, on the mining customs of the Cornish tin miners and early Spanish mining laws introduced into California, the miners established the first governments in the American West. 340 ppgs, 24.99

BLM Field Handbook for Mineral Examiners - Leading mining historian Kerby Jackson introduces us to a classic mining publication on mine evaluation. First published in 1962, this work sheds important light on the techniques of BLM Mineral Examiners to perform validity on mining claims. 132 ppgs, 10.99

Six Months In The Gold Mines During The California Gold Rush - Unavailable since 1850, this important work is a first hand account of one "49'ers" personal experience during the great California Gold Rush, shedding important light on one of the most exciting periods in the history of not only California, but also the world. Compiled from journals written between 1847 and 1849 by E. Gould Buffum, a native of New York, "Six Months In The Gold Mines During The California Gold Rush" offers a rare look into the day to day lives of the people who came to California to work in her gold mines when the state was still a great frontier. **8.5" X 11", 290 ppgs. Retail Price: $19.99**

The Discovery of Gold in Australia - **First published in 1852, it has been unavailable since those days and sheds important light on Australia's gold mining history. Included are rare communications between British agents and the British Crown when gold was first discovered in Australia in 1851. This rare text contains hard to find details on Australia's first mining camps and Britain's early attempts to provide for the orderly regulation of gold mines in that part of the world. Also of interest are hard to find extracts of articles that appeared in the early colonial newspapers that did their best to report on Australia's gold rush as it took place.**
102 ppgs, 10.99